ASEAN 2.0

The **ASEAN Studies Centre** of the Institute of Southeast Asian Studies in Singapore is devoted to working on issues that pertain to the Association of Southeast Asian Nations as an institution and a process, as distinct from the broader concerns of the Institute with respect to Southeast Asia.

Through research, conferences, consultations, and publications, the Centre seeks to illuminate ways of promoting ASEAN's purposes — political solidarity, economic integration and regional cooperation — and the obstacles on the path to achieving them. Through its studies, the Centre offers a measure of intellectual support to the ASEAN member-countries and the ASEAN Secretariat in building the ASEAN Community, with its political/security, economic and socio-cultural pillars. The Centre aims to conduct studies and make policy recommendations on issues and events that call for collective ASEAN actions and responses.

The Centre seeks to work together with other intellectual centres, institutes, think-tanks, foundations, universities, international and regional organizations, government agencies, and non-governmental organizations that have similar interests and objectives, as well as with individual scholars and the ASEAN Secretariat.

The **Institute of Southeast Asian Studies (ISEAS)** was established as an autonomous organization in 1968. It is a regional centre dedicated to the study of socio-political, security and economic trends and developments in Southeast Asia and its wider geostrategic and economic environment. The Institute's research programmes are the Regional Economic Studies (RES, including ASEAN and APEC), Regional Strategic and Political Studies (RSPS), and Regional Social and Cultural Studies (RSCS).

ISEAS Publishing, an established academic press, has issued more than 2,000 books and journals. It is the largest scholarly publisher of research about Southeast Asia from within the region. ISEAS Publishing works with many other academic and trade publishers and distributors to disseminate important research and analyses from and about Southeast Asia to the rest of the world.

**ASEAN
Studies Centre**
Institute of Southeast Asian Studies

Report No. 13

ASEAN 2.0
ICT, Governance and Community in Southeast Asia

EMMANUEL C. LALLANA

INSTITUTE OF SOUTHEAST ASIAN STUDIES
Singapore

First published in Singapore in 2012 by
ISEAS Publishing
Institute of Southeast Asian Studies
30 Heng Mui Keng Terrace
Pasir Panjang
Singapore 119614
E-mail: publish@iseas.edu.sg
Website: bookshop.iseas.edu.sg

The responsibility for facts and opinions in this publication rests exclusively with the author and his interpretations do not necessarily reflect the views or the policy of the publisher or its supporters.

ISEAS Library Cataloguing-in-Publication Data

Lallana, Emmanuel C.
ASEAN 2.0 : ICT, governance and community in Southeast Asia.
1. Telecommunication policy—Southeast Asia.
2. Information technology—Southeast Asia.
I. Title.
HE8380.8 Z5L19   2012

ISBN 978-981-4345-28-6 (soft cover)
ISBN 978-981-4345-29-3 (E-book PDF)

Typeset by Superskill Graphics Pte Ltd
Printed in Singapore by Markono Print Media Pte Ltd

"The very title of this book embodies the concept of flexibility and learning from errors: in the software business, "Release 1.0" is the first commercial version of a new product, following test versions usually called 0.5, 0.8, 0.9, 0.91, 0.92. It is fresh and new, the realization of the hopes and dreams of its developers. It embodies new ideas and it is supposed to be perfect.

Usually the vendor comes out with Release 1.1 a few months later, fixing unexpected bugs and tidying up loose ends.

If the product succeeds, the vendor launches Release 2.0 a year or so later. Release 2.0 is a total rewrite, hammered out by older, wiser programmers with feedback from thousands of tough-minded, sceptical users. Release 2.0 is supposed to be perfect, but usually Release 2.1 comes out a few months later."

Ester Dyson
Author of *Release 2.0: A Design for Living in the Digital Age*

CONTENTS

PREFACE

ASEAN 2.0 is an effort to address my two interests: ICT and global governance. ICT — with its ability to overcome distance and time — should be the ideal tool for enhancing collaboration in creating global rules. Yet, ICT is still largely absent in the current literature on increasing stakeholder participation in global governance.

This study is one of eight projects being undertaken by the Pan Asia Network for eGovernance (PANeGOV). PANeGOV is a network of scholars from South and Southeast Asia and is funded by Canada's International Development Research Centre (IDRC). The ASEAN Studies Centre, Institute of Southeast Asian Studies (ASC-ISEAS), Singapore, also supported the conduct of this study. With the ASC-ISEAS behind this project, I had the pleasure of getting another opportunity to work with former ASEAN secretary general and current ASC-ISEAS head, Ambassador Rodolfo C. Severino. In the mid-1990s, I attended ASEAN Senior Official Meetings (SOMs) as a member of the Philippine delegation led by Ambassador Severino.

While my name appears as the author of the study, any text is hardly the product of one person alone. As noted by cultural critic Roland Barthes in *Death of the Author*, "the text is a tissue of quotations drawn from the innumerable centres of culture". The brief description of the research and writing process that follows will illustrate the collaborations involved in producing this study.

With the help of Mina Peralta, who served as research associate for this study, a first draft was completed at the end of January

2010. Ambassador Severino and Dr Pavin Chachavalpongpun, lead researcher, Political and Strategic Affairs, ASC-ISEAS, read and commented on this draft of the study.

A revised first draft was then made and was reviewed by the following:

- Dr Aileen Baviera, Professor, Asian Studies Centre, University of the Philippines — Diliman, the Philippines;
- Hans Bayaborda, Business Development Director (ASEAN and India), Oracle Corporation, the Philippines;
- Seow Hiong Goh, Executive Director, Global Policy and Government Affairs for Asia Pacific, Cisco, Singapore;
- Ramon L. Jocson, Vice-President, IBM Global Technology Services, Asia Pacific, Singapore;
- Dr Maria Consuelo Ortuoste, Assistant Professor, Department of Political Science, California State University East Bay, California, the United States;
- Virgilio L. Pena, former Chairman, Commission on Information and Communications Technology, Republic of the Philippines; and
- Roberto R. Romulo, former Secretary of Foreign Affairs, Republic of the Philippines.

A new version of the study was produced at the end of March 2010. This was then presented to a series of roundtable discussions (RTD) in all ten ASEAN capitals. The roundtable discussions were conducted over a span of six weeks, from 2 April to 13 May 2010. Below is the detailed schedule of the RTD and our co-hosts:

April 2   –   Yangon RTD hosted by the Tun Foundation Bank

April 5 – Singapore RTD hosted by ASC-ISEAS

April 12 – Jakarta RTD hosted by The Habibie Centre

April 13 – Meeting with the ASEAN secretary general

April 14 – Bandar Seri Begawan RTD hosted by Brunei Darussalam Institute of Policy and Strategic Studies, MFA

April 19 – Manila RTD jointly hosted by the Foreign Service Institute, Department of Foreign Affairs and Ideacorp

April 30 – Bangkok RTD hosted by the Ministry of Foreign Affairs

May 3 – Phnom Penh RTD hosted by the Cambodian Institute for Cooperation and Peace

May 5 – Vientiane RTD hosted by the National Authority for Science and Technology, Department of Science and Technology

May 7 – Hanoi RTD hosted by the Institute of Southeast Asian Studies, Vietnam Academy of Social Sciences

May 13 – Kuala Lumpur RTD hosted by the Institute of Strategic and International Studies

The comments that were given by the discussants and other participants of the various RTDs informed the final writing of the text.

A report on the roundtable discussion was prepared by Ms Peralta and is included in this study as Annex 2.

It was decided early on that there will be two versions of the results of the study.

The first is a short and straightforward report for policy-makers (reproduced here as Annex 1). We sent copies of this version of the study to responsible officials at the Foreign Ministries and ICT ministries of ASEAN countries. I also presented the results of this study to the ASEAN TELSOM at its July 2010 meeting in Singapore.

The second version — the "scholarly" report — is what you now hold in your hands.

Executive Summary

This study looks at four areas that will be important to the future success of ASEAN:

1. How to use Information and Communications Technology (ICT) to achieve better policy coordination, particularly horizontal and vertical policy coordination;
2. How to use ICT to incorporate non-state actors in regional policy-making, thereby creating a more inclusive regional organization;
3. How to use ICT to develop a regional identity and community; and
4. How to use ICT for ASEAN "network management", which involves determining the purpose of the network, as well as establishing all the activities and operations to achieve that purpose.

At present, ICT use in the ASEAN organization is largely limited to a web browser and office productivity tools, including email, word processing, and spreadsheets. This study presents several recommendations for ASEAN to take full advantage of the benefits that ICT use offers.

It is argued in this study that ASEAN should examine new *software-based conferencing solutions* to complement existing face-to-face ASEAN meetings. Moreover, ASEAN can use *collaborative software* to respond to the challenges of horizontal and vertical policy coordination within the organization.

ICT can also help ASEAN achieve its goal of becoming an inclusive institution of regional governance. *eParticipation* — the use of ICT for enabling appropriate stakeholder participation in decision making — can enhance stakeholder participation in creating and strengthening ASEAN processes and policies. For instance, the Senior Official Meeting (SOM)'s engagement with ASEAN Institutes of Strategic International Studies network (ASEAN-ISIS) can be bolstered using ICT.

ASEAN should also consider online engagement with "Entities Associated with ASEAN" — approximately seventy-two organizations that support the purposes and principles of the ASEAN Charter.

In terms of promoting inclusive regional governance, it is suggested that ASEAN consider the use of *wikis* — web pages that anyone with permission can create or edit — as well as *blogs* — shared online journals also called web logs. Wikis and blogs may be used to widen participation in ASEAN policy development and strengthen people-to-people exchange.

With regard to the promotion and creation of a regional identity, ASEAN can use *social network sites* (SNS) to enable the citizens of member states to "imagine" an ASEAN community. ASEAN should leverage the power and appeal of SNS with the aim of developing an ASEAN identity among the youth sector of its member countries.

This study also recommends the appointment of an ASEAN Chief Information Officer (CIO). It is argued that the ASEAN CIO's most urgent task is to establish the foundation for electronic interaction with the following aims: 1) to allow ASEAN and all its stakeholders to share information in real time; and 2) to deploy an electronic coordination mechanism within ASEAN, as well as between ASEAN and its partners and stakeholders.

In closing, this study argues that the successful use of ICT by ASEAN is not merely dependent on, and limited to, correct technology choices or the appointment of a competent CIO. ASEAN officials, particularly those in the highest levels of the institution, must fully recognize ICT as a tool for enhanced governance, effective policy-making, and intensified community building efforts in Southeast Asia.

ASEAN 2.0: ICT, Governance, and Community in Southeast Asia

1. INTRODUCTION

On 20 November 2007, the leaders of the ten states comprising the Association of Southeast Asian Nations (ASEAN) signed the ASEAN Charter. After forty years of existence, the Southeast Asian intergovernmental organization codified the key principles and purposes of the association through the Charter. Then ASEAN Secretary General Ong Keng Yong declared that the Charter serves the organization in three interrelated ways: it formally accords ASEAN its legal personality; it establishes greater institutional accountability and compliance system; and it reinforces the perception of ASEAN as a serious regional player in the Asia-Pacific region.[1]

Signed with the ASEAN Charter are three blueprints that serve as the pillars of the Association towards building and creating an ASEAN community. The first pillar is the ASEAN Economic Community (AEC) blueprint, with the goal of forging economic integration within the region by 2015. The AEC is part of the Association's efforts to "modernise the group's commerce and economic systems to facilitate a better flow of people and goods between countries".[2] The AEC anticipates the following: a single market and production base; a highly competitive economic region; a region of equitable economic development, and; a region fully integrated into the global economy.

The second pillar of the ASEAN community is the ASEAN Political Security Community (APSC) blueprint. The APSC

Blueprint envisages ASEAN as a rules-based community of shared values and norms; a cohesive, peaceful, stable, and resilient region with shared responsibility for comprehensive security; as well as a dynamic and outward-looking region in an increasingly integrated and interdependent world.[3] With the APSC, the members of the community pledge to rely exclusively on peaceful processes in the settlement of intraregional differences. The APSC recognizes that each member country's political security is fundamentally linked to those of the other member countries, as the community is bound by geographic location, common vision, and objectives.

The ASEAN Socio-Cultural Community (ASCC) blueprint is the third pillar of the vision to create an ASEAN community. The ASCC blueprint aims to realize a people-oriented and socially responsible community. It seeks to forge a common identity and build an inclusive, caring, and sharing society where the well-being, livelihood, and welfare of the peoples of the region are enhanced. The ASCC is focused on nurturing the human, cultural, and natural resources of the region for sustained development in a harmonious and people-oriented ASEAN.

But even with its clear goals, the new ASEAN Charter is not without critics. Its lack of mechanisms to ensure the compliance of its member countries with ASEAN obligations, including policies related to issues of democracy, human rights and fundamental freedom, social justice, the rule of law, and good governance have been pointed out. Others are wary of the terms that vest the final powers of decision in the ASEAN Summit, fearing that its decisions would inevitably be politically driven.[4]

Yet, for better or for worse, all agree that with the ASEAN Charter and the blueprints for ASEAN community, the intergovernmental organization has been transformed. ASEAN 2.0 has emerged.

2. INSTITUTIONS, NETWORKS, ICT

The emergence of ASEAN 2.0 can best be appreciated through the lenses of international institutions. This study looks at the institutional design of ASEAN 2.0 as part of the broader effort to understand the linkage between the design features of regional institutions and regional integration and community building.[5] It also gives special attention to the constitutive role of Information and Communications Technology (ICT) in the institutional design of a network organization.

Institutions

International institutions are "explicit arrangements, negotiated among international actors that prescribe, proscribe, and/or authorise behaviour".[6] The term "institutional design" is understood as "those formal and informal rules and organisational features that constitute the institution and that functions as either the constraints on actor choice or the bare bones of the social environment within which agents interact or both".[7]

For Klijn and Koppenjan, "Institutional design is aimed at deliberate changes in institutional characteristics of networks."[8] They are wilful activities that are not tied to "(re)interpretations of actors (creating gradually different understandings of the rules) or more or less conscious ignoring or changing the application of rules".[9] Institutional design "refers both to the *activity* of trying to change the institutional features of policy networks, as [well as] to the *content* of the institutional change that is aimed for". (italics in the original)[10]

Although actors consciously (and rationally) change the design of institutions (as ASEAN did in signing and implementing the Charter), institutional designs are not "rational designs". Rather they are:

the result of the process of pushing and pulling between the parties involved. Policy assumptions about the effectiveness of institutional designs play a role, but so do the power relations between conflicting coalitions.[11]

The actual "shape" of the design of ASEAN 2.0 will be determined by a series of actions and activities by designated actors, as well as the interplay of these actions and activities.

Networks

A specific feature of ASEAN as an institution is that it is a network organization.

"Networks", according to Lawrence J. O. Toole Jr., "are structures of interdependence involving multiple organisations or parts thereof ..."[12] Similarly, for Manuel Castells, a network is "a set of interconnecting nodes that de-centre performance and share decision-making".[13] "In social theoretical terms", according to Mark Considine, networks are "defined by a qualitatively different means by which actors change their work, create and sustain new forms of technologically assisted production and reproduction, and alter the participation and identity of themselves and others".[14]

ASEAN, as an intergovernmental organization, exhibits the characteristics that Charlotte Streck suggests all networks share:

- Networks are based on informal arrangements instead of legally binding agreements;
- Cooperation in networks is based on trust and not on enforceable obligations;
- Cooperation in networks is voluntary in its nature;
- Networks are open to allow other partners/actors to join;

- The partners in a network bring different resources and assets to the table;
- Networks are loosely structured; and
- Networks evolve over time.[15]

In terms of governance, a network employs a distinct form of coordinating its activity which is very different from a hierarchy and market.[16] Unlike a hierarchy with its "pyramid of offices" that is coordinated through command, network organizations operate both horizontally and vertically and "achieve coordination through mutual adjustment".[17] While networks and markets are both "exchange relationships", "(m)arket exchange implies relatively discrete and/or impersonal exchange relationships, (while) network exchange is more diffuse and/or more social".[18]

Reflecting upon the experience of the European Union (EU), Andrew Jordan and Adriaan Schout suggest that:

> ... when properly harnessed (i.e. managed) networks are, in some situations, capable of providing an alternative means of fulfilling complex policy goals to hierarchy. This is because in modern policy systems (particularly polycentric ones like the EU), central bodies have the diminished capacity to exert hierarchical authority, and should instead be viewed as a participant — albeit a special one — in inter-organisational networks.[19]

When we combine the aforementioned definitions, ASEAN is best understood as a network organization. To be more precise, ASEAN is a network of policy networks.

Each sector meeting (AMM, AEM, ASEAN TELMIN, etc.) is a policy network composed of the relevant ministries of ASEAN member countries. All these policy networks are tied together in an overarching network that is ASEAN. When viewed as a network organization, it becomes apparent that ASEAN operates much like the Internet.

While many have asserted the superiority of the network organization, the traditional weakness of these organizations lies in their coordination.[20] In this study, it will be argued that ASEAN's success will be partly hinged on its ability to use ICT in its operations, particularly in coordination. ICT is an important tool not only because it facilitates coordination, but also because it changes the ways coordination is carried out.[21]

Thus, aside from contributing to an understanding of ASEAN's institutional design in the context of networks, this study also hopes to contribute to the growing literature on ICT and governance or eGovernance.

ICT

eGovernance refers to the use of ICT in the "fundamental organisational processes characteristic of groups and wider social structures and comprising the institutions or mechanisms which enable them to coexist, whether formal or informal, explicit or implicit".[22] More specifically, eGovernance is the use of ICT by government and other political actors in the domains of administration (public service delivery, regulation, law enforcement, security, improving bureaucratic efficiency) and politics (constitutional procedures and rule making) at the local, national, regional, and global levels.

The above definition of eGovernance recognizes that while the interdependencies among government, private, and non-profit sectors in social steering is inevitable, the state continues to play an important role (the so-called "governance **with** government").

Furthermore, and key to this discussion, this definition underscores the fact that the use of ICT by political actors to improve coordination, arbitration, networking, and regulation (essential steering functions) occurs not only at the nation state level, but at regional (Southeast Asia, South Asia, East Asia, Asia Pacific, or Europe) and global levels.

In its early use, eGovernance was synonymous with eGovernment — the use of ICT to:

> a) enhance the access to and the delivery of Government information and services to the public; and b) to improve the effectiveness, efficiency and/or the quality of Government operations.[23]

Eventually, eGovernance was expanded to mean:

> the use of ICT to create a wealth of new digital connections including: 1) connections within government — permitting 'joined-up thinking'; 2) connections between government and NGOs/citizens — strengthening accountability; 3) connections between government and business/citizens — transforming service delivery; 4) connections within and between NGOs — supporting learning and concerted action; and 5) connections within and between communities — building social and economic development.[24]

More recently, the increasing use of the concept, Government 2.0, has become apparent. What follows is the Australian Government's definition of Government 2.0 that is widely shared:

> [It] is about the use of technology to encourage a more open and transparent form of government, where the public has

a greater role in forming policy and has improved access to government information.[25]

Moreover, for Tapscott and Williams, the extensive use of social media allows government to become "a platform for social innovation".[26] This is because "it provides resources, sets rules, and mediates disputes, but allows citizens, nonprofits, and the private sector to share in the heavy lifting".[27]

It is argued that Government 2.0 is able to realize the promise of collaborative policy-making because it is based on Web 2.0, which is "The second generation of the World Wide Web, especially the movement away from static WebPages to dynamic and shareable content and social networking."[28] It is technology that is able to facilitate creativity, information sharing, and, most notably, collaboration.

A significant example of how technology enables information sharing is the emergence of the "blog" or "web log" — "a type of website usually maintained by an individual with regular entries of commentary, descriptions of events, or other material such as graphics or video".[29] The blog has lowered the entry barrier for public political commentary. Anyone with a penchant to write, is ICT literate, and has Internet access, can set up a blog in minutes and begin sharing ideas with the world shortly thereafter.

While there are many types of blogs, what we are interested in are "political blogs" and the "policy blogs". Political bloggers use their blogs in four ways: 1) as transmission belts or links providers to web sources with no commentary; 2) as "soapboxes" to spout opinions, record a political diary, or to confess; 3) as "mobilisers" to encourage readers to take political action; and 4) as "conversation starters" to elicit feedback, encourage dialogue, and ask questions.[30] On the other hand, "policy bloggers" refer to those who "use their blogs to filter information, to provide expertise, to form networks, to attract attention, to frame arguments, and to exploit windows of opportunity".[31] Policy blogs provide interested citizens with information surrounding policy goals and instruments, and provide a venue for government officials to discuss issues openly. It is also a means for ordinary citizens to share their views with policy-makers — a new channel of communications (and perhaps collaboration) between government and citizens.

The development of "wiki" websites may be cited to illustrate the value of technology to policy-makers. A wiki "is a website that allows multiple users to create, modify and organise web page content in a collaborative manner".[32] There are many kinds of wikis: those interested in governance focus on two specific variants — the "policy wiki" and the wiki as a knowledge management tool.

A policy wiki is:

> a combination of a traditional wiki — that is, a publicly-editable resource similar to Wikipedia — and a public discussion forum, with comments and voting features. In many ways, it's a kind of social-media mash-up aimed at pulling in suggestions from readers and other concerned (citizens) about public policy issues.[33]

A good example of a policy wiki is New Zealand's Police Act wiki. In 2007, when New Zealand's "Police Act of 1958" was being amended, the New Zealand Police Act review team ran a wiki to solicit citizens' input for the new law. The wiki contained the contents of the "Police Act 1958" and allowed anyone to edit it as they would a Wikipedia article. The wiki allowed for broad consultation at low cost and generated new ideas for legislators. New Zealand's efforts at employing a new channel for public participation are a preview of how collaborative technologies incorporate citizens in policy discussions and development.

Apart from information sharing, collaboration, and consultation, wikis are used for knowledge management for at least three reasons: 1) they can disseminate knowledge to domains across time, distance and organizations; 2) they can be integrated as part of a work process in order to facilitate knowledge sharing and system utilization; and 3) they provide a dynamic system for knowledge capture that is able to evolve with changing needs.[34]

The *Diplopedia* is an example of a successful use of the wiki for knowledge management in government. The Diplopedia is the "U.S. State Department's internal knowledge-sharing platform — an open-source wiki platform of 'how-to' knowledge for America's diplomatic corps".[35] It was created to address the problem of knowledge loss following the transfer of a Foreign Service Officer (FSO) from one post to another. Prior to the *Diplopedia*, there was no sufficient means that allowed a transferred officer to share his/her accumulated knowledge to his/her successor.

Launched in 1996 with ten "seedling" articles, *Diplopedia* contributors have added more than 10,000 articles to the website by January 2010.[36] Usage has also grown significantly — from an average daily visit of 500–800 in August 2008 to more than 2,000 visits by February 2010. *Diplopedia* is now "a reference and starting point for all topics of interest to the Department (of State) and U.S. Government (USG) foreign affairs community".[37]

An increasing number of national governments have already embraced Government 2.0, including Australia, Canada, the Netherlands, New Zealand, the United Kingdom, and the United States.[38] Local governments have also launched Government 2.0 initiatives. Examples include "Melbourne, Australia's *FutureMelbourne*,[39] *localocracy* of Amherst (Massachusetts) in the USA",[40] and *openly local* of the Aberdeen City Council (the United Kingdom).

The above examples show the benefits of eGovernance, including information sharing and public consultation. Despite these gains, Beth Simon Noveck laments:

> Finally, the digital environment offers new ways to engage in the public exchange of reason. With new tools, people can "speak" through shared maps and diagrams rather than meetings. Competing proposals, using computer-driven algorithms and prediction markets, can evolve. Policy simulations using graphic technology can be created. Social networking tools enable collaborative making, doing, crafting and creating. Yet most of the work at the intersection of technology and democracy has focused on how to create demographically representative conversations. The focus is on deliberation, not collaboration, on talk instead of action; on information and not decision-making.[41]

Furthermore, little effort has been made for the purpose of using ICT to enhance what John Dryzek calls, "deliberative global politics".[42] This is unfortunate since ICT is probably the best tool to overcome the problem of stakeholder participation in global governance.

This study is a contribution to "correct" these deficiencies related to the fusion of technology and governance. It is also an extended argument on how to incorporate ICT in ASEAN's emergent institutional design.[43] Specifically, this study will look at four areas that will be important to the future of ASEAN:

- How to use ICT to achieve better policy coordination — particularly horizontal and vertical policy coordination;

- How to use ICT to incorporate non-state actors in regional policy-making thereby creating a more inclusive regional organization;
- How to use ICT to develop a regional identity and community; and
- How to use ICT for network management (which involves not only determining the purpose of the network, but also establishing all the activities and operations for achieving that purpose).

3. FROM ASEAN 1.0 TO ASEAN 2.0

From its inception, ASEAN can be viewed as a governance network which, according to Sørensen and Torfing, can be characterized as:

- A relatively stable horizontal articulation of interdependent, but operationally autonomous actors;
- An organization with actors who interact through negotiations;
- An organization whose negotiations take place within a regulative, normative, cognitive, and imaginary framework;
- An organization that, to a certain extent, is self-regulating; and
- An organization which contributes to the production of public purpose within or across particular policy areas.[44]

Moreover, Hadi Soesastro's description of ASEAN "as a loose form of inter-governmental cooperation that accords highest priority to the preservation of national sovereignty" allows us to view ASEAN from a network perspective.[45]

One reason networks are formed is to "enhance horizontal coordination", and they are partly held together "by the anticipated gains from resource pooling and joint action and partly by the development of mutual trust that helps to overcome collective action problems".[46] Take for instance the reasons given by Thanat Khoman for the founding of ASEAN:

The most important of them was the fact that, with the withdrawal of the colonial powers, there would have been a power vacuum which could have attracted outsiders to step in for political gains. As the colonial masters had discouraged any form of intra-regional contact, the idea of neighbours working together in a joint effort was thus to be encouraged.

Secondly, as many of us knew from experience, especially with the Southeast Asia Treaty Organisation (SEATO), cooperation among disparate members located in distant lands could be ineffective. We had therefore to strive to build cooperation among those who lived close to one another and shared common interests.

Thirdly, the need to join forces became imperative for the Southeast Asian countries in order to be heard and to be effective. This was the truth that we sadly had to learn. The motivation for our efforts to band together was thus to strengthen our position and protect ourselves against Big Power rivalry.

Finally, it is common knowledge that cooperation and ultimately integration serve the interests of all — something that individual efforts can never achieve.[47]

But the development of any organization takes time. Forty years after the establishment of ASEAN, Carolina Hernandez characterized ASEAN's experience at institution building as "painfully slow and incremental at best".[48]

For Sridahan, ASEAN has gone through three distinct phases in its institutional evolution. Phase 1 (or what we will call ASEAN 1.0) covers ASEAN's development as an organization from 1967 to 1975. This period was "marked by a very loose, disjointed structure which was dominated by a few committees headed by officials".[49] In a way, the preference for a loose organization is embedded in the founding document of ASEAN itself.

According to the ASEAN Secretariat:

> The two-page Bangkok Declaration not only contains the rationale for the establishment of ASEAN and its specific objectives. It represents the organisation's modus operandi of building on small steps, voluntary, and informal arrangements towards more binding and institutionalised agreements.[50]

Another feature of ASEAN 1.0 is the dominance of ASEAN foreign ministers. As noted by Soesastro:

> ASEAN was founded and formulated by foreign ministers, and the meeting of foreign ministers, known as the AMM, was designed to become the central institution of ASEAN. The day-to-day work of the AMM is carried out by a Standing Committee, headed by a foreign minister from a member country on a rotational basis, and not by a regional body.[51]

Phase 2 of the ASEAN institutional evolution (or ASEAN 1.1 in this author's version) is from 1976 to the early 1990s.[52] A key feature of this period is the first Summit of ASEAN's heads of government/state. This summit produced the "Declaration of ASEAN Concord and the Treaty of Amity and Cooperation".

The "Declaration of ASEAN Concord" defined the programme of action (in the political, economic, social, cultural and information, security, and improvement of ASEAN's machinery) as a framework for ASEAN cooperation. The "Treaty of Amity and Cooperation in Southeast Asia" (TAC) builds on the Bangkok Declaration and enumerated the Association's principles that govern state conduct and relations within the region:

- Mutual respect for the independence, sovereignty, equality, territorial integrity, and national identity of all nations;
- The right of every state to lead its national existence free from external interference, subversion or coercion;
- Non-interference in the internal affairs of one another;
- Settlement of differences or disputes in a peaceful manner;
- Renunciation of the threat or use of force; and
- Effective cooperation among themselves.[53]

ASEAN 1.1 also saw the emergence of the ASEAN Economic Ministers (AEM) as the other power centre of the regional organization. The Senior Economic Officials Meeting (SEOM) was created to support the AEM and was responsible for overseeing the organization's economic cooperation activities. This phase also saw the creation of the ASEAN Secretariat headed by a secretary general.

The third phase of ASEAN's institutional evolution (or ASEAN 1.2) moves on from the ASEAN Summit in 1992 to the adoption of the ASEAN Charter in 2007. The most significant change was the institutionalization of the ASEAN Summit. It was also during ASEAN 1.2 that the position of the secretary general was upgraded from being the head of the ASEAN Secretariat to becoming the secretary general of ASEAN. The 1992 ASEAN Summit was also the occasion when the ASEAN Free Trade Agreement (AFTA) was signed. Consequently, ASEAN's economic perspective transformed from economic cooperation to economic integration.

A common feature of ASEAN's institutional design in its first forty years was "exclusive executive multilateralism".[54] This means that as an organization, ASEAN is "characterised by non-public negotiations and bargaining between national government representatives which are consciously isolated from public scrutiny or participation".[55]

In its first forty years, the "ASEAN Way" was a constitutive feature of ASEAN. According to Soesastro, the "ASEAN Way" had the following key elements:

- Preference for weak institution;
- Emphasis on confidence building;
- Intensive process;
- Strong intergovernmental presence;
- Consensual decision making; and
- Sovereignty enhancing.[56]

The "ASEAN Way" was not only how ASEAN conducted its business, but was also offered as a norm in a Pacific-wide security cooperation initiative. For ASEAN member states, the "ASEAN

Way" helped "ensure that national governments retained sufficient autonomy to determine domestic policy on key issues in line with its domestically derived priorities rather than be compelled to follow an externally mandated policy agenda".[57] In this regard, it is also important to remember that:

> The ASEAN Way itself stemmed not so much from deeply rooted cultural sources like Javanese or other else but from incremental socialisation. It emerged not merely from the principles of interstate relations agreed to by the founders of ASEAN, but also from a subsequent and long-term process of interaction and adjustment. Hence, in the case of ASEAN, it is evident that culture created norms and norms also created culture.[58]

In his review of ASEAN's first forty years, Acharya noted that "The 'ASEAN way' of informal networking has ... trumped efforts to institutionalise cooperation."[59]

ASEAN 2.0

The move to ASEAN 2.0 reflects changes in the global order. As globalization resulted in the deepening interdependence of nations, there was a need for ASEAN to adapt and reposition itself in order to meet these fundamental changes. According to Ali Alatas, the rise of globalization led to the "realisation" that "the way ASEAN has been functioning over the years would not suffice any longer".[60]

In terms of institutionalization, ASEAN 2.0 entails:

- An established legal personality for the organization;

- The ASEAN Summit as the supreme policy-making body meeting twice a year;
- The presence of the ASEAN Coordinating Council and Community Councils to implement the decisions of the Summit;
- The presence of the ASEAN Secretariat as the administrative body to monitor the progress of the implementation of ASEAN decisions;
- The presence of disputes resolution instruments;
- The presence of provisions to advance the goal of community building, including Articles 35–40 that established an ASEAN motto, flag, anthem, and an official day.[61]

By the end of 2009, Tommy Koh had identified "four substantive achievements" since the Charter was enforced:

1. The establishment of the Committee of Permanent Representatives (CPR);
2. The establishment of the ASEAN Intergovernmental Commission on Human Rights (AICHR);
3. The adoption of the agreement on privileges and immunities; and,
4. The significant strengthening of ASEAN's economic pillar because the goal of achieving a single market and production base by 2015 is on track.[62]

As will be elaborated subsequently, these changes in institutional design are aimed at "strengthening network interactions" with the hope to "influence the interactions between actors in a sustainable way".[63]

4. ICT IN HORIZONTAL POLICY COORDINATION IN ASEAN

States have agreed to coordinate policy for a variety of reasons. In the case of international macro-economic policy from 1945 to 1988, policy-makers had a "preference for coordination based around targeting of exchange rates".[64] Exchange rates were targeted because: 1) exchange rates stability is important in itself; 2) it helps "dampen fads and speculative bubbles in foreign exchange markets"; and 3) it serves as a discipline on the monetary and fiscal policies of the participating countries.[65]

According to Pelkonen, Teräväinen, and Waltari, *vertical policy coordination* "refers to managing relationships between various levels of government and proceeding from priority-setting to policy implementation". On the other hand, *horizontal policy coordination* refers to the "management of interdependent policies across the state administration".[66] In this study, vertical policy coordination in ASEAN is synonymous with sectoral policy-making (i.e., the policy-making activities of the various sectoral bodies and processes such as the ASEAN Telecommunications and IT Ministers Meeting (TELMIN) and the ASEAN Meeting of Energy Ministers (AMEM)). Horizontal policy coordination refers to the procedures and offices that are established to ensure that the overall goal of community building is served and that there are minimal overlaps or inconsistencies in the output of the various sectoral bodies and processes.

ASEAN has created new institutions to ensure the implementation of horizontal policy coordination as reflected in its new charter.

One such institution is the ASEAN Coordinating Council, composed of ASEAN foreign ministers, which is specifically tasked to "coordinate the implementation of agreements and decisions of the ASEAN Summit" and "coordinate with the ASEAN Community Councils to enhance policy coherence, efficiency and cooperation among them" (Article 8, Section 2). Like other ASEAN bodies, the ASEAN Coordinating Council is supported by the "relevant senior officials".

Three ASEAN Community Councils (Political-Security, Economic Community, and Socio-Cultural) were also created. These Community Councils are mandated to "coordinate the work of the different sectors under their purview, and on issues which cut across the other Community Councils", (Article 9) among other responsibilities.

There are two other ASEAN bodies relevant to horizontal policy coordination: the ASEAN Secretariat and the Committee of Permanent Representatives to ASEAN.

The Committee of Permanent Representatives (CPR) is composed of member country ambassadors based in Jakarta and is tasked to:

1. Support the work of the ASEAN Community Councils and ASEAN Sectoral Ministerial Bodies;
2. Coordinate with ASEAN National Secretariats and other ASEAN Sectoral Ministerial Bodies;
3. Liaise with the secretary general of ASEAN and the ASEAN Secretariat on all subjects relevant to its work; and,
4. Facilitate ASEAN cooperation with external partners. (Article 12)

The ASEAN Secretariat (Secretariat), headed by the ASEAN secretary general:

facilitates and monitors progress in the implementation of ASEAN agreements and decisions; submits an annual report on the work of ASEAN to the ASEAN Summit; and participates in meetings of the ASEAN Summit, the ASEAN Community Councils, the ASEAN Coordinating Council, and ASEAN Sectoral Ministerial Bodies and other relevant ASEAN meetings. (Article 11) The ASEAN Secretary General is designated as the "Chief Administrative Officer of ASEAN".

According to Tommy Koh, one year after the adoption of the ASEAN Charter, the Committee of Permanent Representatives (CPR) "has begun to function effectively. Working closely with the ASEAN Secretariat, it is coordinating a broad range of issues and ensuring that decisions are taken in a timely manner."[67]

Meanwhile, the ASEAN bodies related to vertical policy coordination fall under the ASEAN Sectoral Ministerial Bodies. These bodies shall:

1. Function in accordance with their respective established mandates;
2. Implement the agreements and decisions of the ASEAN Summit under their respective purview; and
3. Strengthen cooperation in their respective fields in support of ASEAN integration and community building. (Article 10)

The ASEAN National Secretariat (of each member state) serves as a national focal point of ASEAN; coordinates the implementation of ASEAN decisions at the national level; coordinates and supports the national preparations of ASEAN meetings; promote ASEAN identity and awareness at the national level; and contributes to ASEAN community building. (Article 13)

The other bodies created by the ASEAN Charter (i.e. those that were previously identified as having roles in horizontal policy coordination) also play a role in vertical policy coordination in that all decisions reached by all ASEAN bodies must be consistent with, and affirmed by, the "highest" decision-making authority — the ASEAN Summit.

A key challenge to ASEAN 2.0 is how to enhance the "coordination capacities" — mechanisms that facilitate coordination within networks of interdependent actors — of these new (and revitalized) bodies. In general, ASEAN can enhance coordination capacities by helping member states to enhance the exchange information amongst them, identifying issues requiring coordinated solutions; as well as arbitrating when conflicts cannot be settled informally by the participants.[68]

At the nation state level, governments normally have three tools for achieving horizontal and vertical policy coordination, which are:

- Political tools — such as government meetings, government/ministerial committees;
- Administrative tools — such as inter-ministerial consultations, meeting of administrative heads of ministries (to coordinate policies and legislation before they reach final decision at the political level); and
- Procedural tools — including rules of procedure, annual planning system, etc.[69]

In order to enhance horizontal and vertical policy coordination in ASEAN, there is a need to implement procedural innovations. For Ian Peach, leadership in procedural innovations is necessary to implement horizontal policy-making and policy implementation.[70]

These innovations include: a results-based reporting and accountability process; reduced demand for procedural reporting and centralized decision making on policy implementation; incentives in the budget for departments to cooperate; and significant recognition and performance bonuses to senior management and staff who make interdepartmental cooperation work effectively.[71] One way to enact procedural innovation in ASEAN is to integrate ICT in horizontal and vertical policy coordination.

For intergovernmental organizations such as ASEAN, it has been noted that transaction costs related to coordination issues increase "with the increase in the number of actors that participate in negotiations".[72] Fortunately, studies on the use of ICT in the industry show that extensive use of ICT:

- reduces the unit of cost of communicating and reacting to information, thereby reducing costs of explicit coordination;
- increases information availability, thereby reducing operations risk; and,
- enhances monitoring capability of ICT and facilitates the monitoring of access to and use of information and expertise, thereby reducing an opportunism risk.[73]

A specific ICT application that ASEAN can use to reduce transaction costs in horizontal and vertical policy cooperation is "collaborative software" or "groupware". Collaborative software is an application (programme) that supports individuals and their interactions with other individuals in the decision-making processes. "The design intent of collaborative software (groupware) is to transform the way documents and rich media are shared in order to enable more effective team collaboration."[74]

Software such as email, calendaring, chat, and web meetings belong to this category of collaborative/groupware. In terms of levels of collaboration, collaborative software/groupware can fall in any or all of these subcategories: 1) communication tools; 2) conferencing tools; and 3) collaborative management (or coordination) tools.

Communication tools, which include email, chat, websites, and blogs, "facilitate the sharing of information between people and hence the sharing of information".[75] Most ASEAN member states and the ASEAN Secretariat are using communication tools.

Conferencing tools enable individuals in various locations to meet online. Among the common features of conferencing tools are: screen-sharing capabilities; multiple presenters; drawing and annotation tools; whiteboard; text chat (or instant messaging); teleconferencing; Voice over Internet Protocol or VoIP; and videoconferencing.[76] At present, conferencing tools are IP and web-based thereby making the cost of conferencing very low (almost on par with the cost of the Internet connection itself). Furthermore, choosing from a wide range of conferencing tools has been made easier by numerous published articles aimed at helping organizations choose the best web conferencing tools for their requirements.[77]

Collaborative management (coordination) tools are those that facilitate and manage group activities.[78] Examples include:

- Electronic calendars to schedule events and automatically notify and remind group members of these events;
- Project management systems to schedule, track, and chart the steps in a project as it is being completed;
- Workflow systems for collaborative management of tasks and documents within a knowledge-based business process; and

- Knowledge management systems that collect, organize, manage, and share various forms of information.

A system called Drupal is an example of a collaborative management tool. It is an open source, content management framework that allows document management, the creation and maintenance of websites, attachments, forums, photos, social profiles, and other collaboration tools. According to Wikipedia, "It is used as a back-end system for many different types of websites, ranging from small personal blogs to Enterprise 2.0 collaboration and knowledge management uses to large corporate and political sites, including <www.whitehouse.gov>."[79]

The experience of organizations with collaborative management tools affirms the importance of such software. Oracle relies on its own enterprise collaboration software called Beehive to coordinate its global and regional activities. Beehive is billed as "the only unified collaboration solution built for the enterprise".[80] It has three main components: 1) Enterprise Messaging — which includes email, calendar, address book, and task management accessible via Microsoft Outlook, Zimbra Web Client, and a selection of mobile phones; 2) Team Collaboration — ICT-enabled Team workspaces with document library, team wiki, team calendar, team announcements, RSS, and contextual search; and 3) Synchronous Collaboration — including Web conferencing and VOIP audio conferencing. Another benefit of the use of Beehive (and its predecessor, Oracle Collaboration Suite) is reduced travel for Oracle staff.

Among the other collaboration tools that ASEAN could use is the "social enterprise software".[81] Such software makes use of social networking software within the enterprise. It marries "consumer-style microblogging, social networking, and related capabilities with the security and management that (enterprise) IT and legal departments demand".[82]

In social enterprise software, wikis are maintained within the secured enterprise firewalls. This is a new possibility for employing wiki for policy work. For Noveck, a policy wiki is a "website where the goal, such as drafting an air quality criteria document, is described and broken down into specific tasks, which small groups of people can elect to undertake — experts and non-experts — alike".[83] These activities include: drafting and posting background research materials relevant to policy issues; inviting experts and other participants to join an advisory network; researching the claims in documents submitted; commenting on and editing particular provisions already drafted; vetting, evaluating, and rating the comments of others; summarizing and translating texts; analysing positions of stakeholders and interested parties; and creating visualizations (diagrams, charts, and illustrations) to reflect and represent the draft.[84]

If policy wikis are used by working groups, senior officials and ministerial meetings before their respective face-to-face meetings, significant reduction in the length of these meetings can be expected. The extensive use of policy wikis may also lead to less travel for all concerned. Thus, such software tools lead to the added benefit of reducing the carbon footprint of ASEAN diplomats.

5. ICT AND INCLUSIVE REGIONALISM

It has been observed that "One of the most profound trends in global governance over the past two decades is the growing extent to which international institutions offer mechanisms for the participation of transnational actors."[85] These "inclusive institutions of global governance" offer private actors from the business sector and/or civil society with the possibility of formal participation, decision-making in the policy-making process.[86]

As an organization, one of the primary objectives of ASEAN is the institution of an inclusive kind of governance.

During the 38th anniversary of ASEAN, Indonesia's President Yudhoyono underscored the need to "enlist the people into the cycle of planning, implementation, monitoring and re-planning of programs and projects".[87] While acknowledging the important role played by the business community and academia in the ASEAN process, he acknowledged that "the perspectives from the boardrooms and from the groves of academia are not the same as the view from the grassroots. And that grassroots view can make a difference."[88] He concluded his talk by suggesting that "We can and should empower the people to become co-authors if not the principal authors of their own development."[89]

Soesastro argued that second track networks and processes, which include research and strategic studies by institutes and other non-governmental participants, is an important institutional feature of ASEAN.[90] In fact, three non-state actors (the business community, research community, and non-government organizations (NGOs)/civil society organizations (CSOs)) have had varying degrees of success in terms of participating in ASEAN processes.

In the 1990s, the ASEAN Chamber of Commerce and Industry (ASEAN-CCI) was the private sector voice in ASEAN.[91] The ASEAN-CCI is comprised of national level chambers of commerce and industry from each of the ASEAN member countries. ASEAN-CCI representatives attended Senior Economic Officials' Meetings (SEOMs) and ASEAN Economic Ministers' (AEM) meetings and have been credited for helping design the ASEAN Free Trade Agreement (AFTA). But the ASEAN-CCI was eclipsed in 2001 when the ASEAN leaders established the ASEAN Business Advisory Council (ABAC). ABAC is composed of up to three businessmen from each member state and has since provided private-sector input into the various ASEAN programmes on economic integration and economic development.

The most successful non-government entity to engage with ASEAN is the network of research organizations collectively known as the ASEAN Institutes for Strategic and International Studies (ASEAN-ISIS). ASEAN-ISIS, which was initiated in 1984 but formally launched only in 1988, comprised Indonesia, Malaysia, the Philippines, Thailand, and Singapore as original members.[92] Initially, each ASEAN-ISIS member worked through its respective governments by submitting ASEAN-ISIS memoranda on various policy issues. Beginning in 1993, ASEAN-ISIS has been directly engaging ASEAN senior officials during the Senior Officials' Meetings (SOM) to discuss outstanding political-security issues facing the region. As noted by Hernandez:

> The degree to which ASEAN-ISIS has succeeded in influencing the foreign policy-making bodies of ASEAN and several other governments in the Asia Pacific and the extent it played the role of track two diplomacy may be seen in the institutionalisation of meetings between ASEAN-ISIS and ASEAN policy making structures such as the ASEAN-SOM, in the adoption of an overwhelming majority of its policy recommendations by ASEAN, in the policy research it has been commissioned to undertake by the ASEAN Secretariat, and in its initiation of and participation in policy discussions with the foreign policy community in many countries in the region, including Japan, Australia, Canada, and the US.[93]

In the 1980s, ASEAN — through the communiqués of the ASEAN Ministerial Meetings (AMM), began to recognize the role of NGOs and started calling for close affiliation with relevant organizations. In June 1986, the ASEAN Standing Committee (ASC) adopted the "Guidelines on ASEAN's Relations with Nongovernment Organisations".[94] This document was revised by the ASC in January 2006 and is now called "Guidelines on ASEAN's Relations with Civil Society Organisations". The change in title underlined a bigger shift in ASEAN-NGO/CSO relationship.

The most significant initiative from ASEAN to establish links with NGOs was manifested during the creation of the ASEAN Charter, when consultations and discussions were held together with the Asian Human Rights Commission. Prior to this, ASEAN's overtures to the NGO community were seen as limited. However, Ortueste noted that the recognition given by the AMM to NGOs in the farmers' communiqués "was ... only a limited opening. The ASEAN governments still decided who could be acceptable partners."[95] It was also noted that many of the NGOs were registered under the ASEAN guidelines were "professional organizations".

According to a regional human rights organization, "civil society at the regional and national levels began to consider ASEAN as an arena through which to pursue their objectives for the promotion and protection of human rights of ASEAN peoples at the regional level" soon after ASEAN announced the drafting of an ASEAN Charter.[96] Thus the Solidarity for Asian People's Advocacy (SAPA) decided to establish the Task Force on ASEAN and Human Rights (TFAHR) (which eventually consisted of more than seventy NGOs) to coordinate collective initiatives on ASEAN advocacy and actions. In 2006 and 2007, the SAPA TFAHR with the Southeast Asian Committee for Advocacy (SEACA), supported national consultations on ASEAN and its proposed Charter with civil society groups from Myanmar/Burma, Cambodia, Indonesia, Laos, Malaysia, the Philippines, Thailand, and Vietnam.[97] The results of these national consultations were presented at two ASEAN Civil Society Conferences held along with the ASEAN Summit. SAPA TFAHR also submitted recommendations on the three pillars of an ASEAN Community to the ASEAN Eminent Persons Group and the High Level Task Force.

While not completely happy with the end version of the Charter and the regional human rights body that was eventually adopted by ASEAN, CSOs continue to engage with ASEAN. As noted by a Thai commentator:

> One of the biggest disappointments in (2009) was the failure of ASEAN's engagement with the civil society organisations (CSO) based in the region. The idea of interface between the ASEAN leaders and CSO representatives at the 14th and 15th ASEAN Summits was to build up trust and start a long standing process of dialogue. All concerned parties have learned very much to their regret that bringing people at the top to converse with the people at the grassroots level would require better preparations and longer processes of dialogue and consultation. At the recent symposium on stakeholders' involvement in regional organisations hosted by the ASEAN Secretariat in Jakarta, representatives from ASEAN and CSOs sat down again in an informal setting trying to find out what went wrong in the past year and discussing new ways to re-engage each other.[98]

Tommy Koh argues that, "Dialogues between political leaders and civil servants, on the one hand, and representatives of civil society, on the other, have been institutionalised."[99] The challenge then is for ASEAN to explore how to use ICT to enhance its policy dialogue with its stakeholders (i.e., civil society organizations, academia and research institutes, and interested citizens of ASEAN member countries). In this regard, ASEAN could consider initiating procedures that fall under the category — e*Participation*.

e*Participation*, according to Wikipedia, "is the generally accepted term referring to 'ICT-supported participation in processes involved in government and governance'".[100]

A specific form of e*Participation* is *e-rulemaking* or the use of ICT to:

- Help develop government regulations;
- Make rulemaking materials broadly available online, along with tools for searching, analysing, explaining, and managing the information they contain; and
- Enable more effective and diverse public participation.[101]

e-rulemaking is already being practised in the United States. A 2008 report to the U.S. president and Congress noted that:

> more than 170 different rulemaking entities in 15 Cabinet Departments and some independent regulatory commissions are now using a common database for rulemaking documents, a universal docket management interface, and a single public website for viewing proposed rules and accepting on-line comments.[102]

The existing U.S. e-rulemaking system is comprised of three interrelated modules:

1. An electronic repository for digitized versions of rulemaking documents organized in electronic dockets, with associated document management capabilities (FDMS e-docket);
2. A password protected interface through which agencies access the repository (FDMS.gov); and
3. The public interface through which those outside the federal government access publicly available materials in FDMS, and can submit comments on proposed rules (Regulations. gov).[103]

Another interesting U.S. effort at engaging the public in government decision making is the "Peer to Patent" project.[104] According to its main proponent, the goal is to "transform (the Patent Office's) closed, centralised process and construct architecture for open participation that unleashes the 'cognitive surplus' of the scientific and technical community".[105] This project aims to institutionalize a "Community Patent Review" process within the U.S. Patent Office.

The project was initiated to aid the processing of patent applications. The United States Patent and Trademarks Office (USPTO) receive close to 500,000 applications a year. With only 5,500 patent examiners, it is not surprising that they have a backlog of one million applications with 120,000 being processed.[106] But it is not only the number of patents that examiners have to cope with; it is also their complexity of the applications.

The backlog is not due to patent examiners taking too long to review applications. Within only twenty hours, patent examiners are expected to 1) read and digest the meaning of the application's claim, 2) research the relevant literature, 3) apply the legal standards of novelty and non-obviousness to the scientific facts, and 4) write up the decision.[107] Thus, the majority of patent examiners complain that they cannot produce quality work during the time required for them to complete their examination.

To address the situation, computerization and electronic filing of patent applications have already been adopted by the USPTO. However, in 2008, the USPTO union observed:

> ... a monumental system for patent examination is being put into place by automation specialists who seem to spurn information from those whom the system is allegedly designed ... rendering the chances for writing software that meets our needs a near impossibility.[108]

The heart of the Peer to Patent project is a website aimed at soliciting information from the public — information that would help the patent examiner do a more thorough assessment of the claims of patent applicants. The Peer to Patent process is all of five steps:

1) Review and discuss patent applications;
2) Research and find prior art;
3) Upload prior art relevant to claims;
4) Annotate and evaluate all submitted prior art; and
5) "Top ten" prior art references forwarded to USPTO.[109]

These are the results of the Peer to Patent pilot as discussed by Noveck:

> The Peer to Patent pilot was intended to test the hypothesis that groups of distributed, self-selected, nongovernmental experts, coming together via the web, could produce expertise to assist the Patent Office with decision-making. In a little more than a year, the pilot attracted 2,300 volunteer reviewers,

working on eighty-four applications. Of the volunteers at least 365 were actively participating reviewers who posted prior art, participated in discussions, and rated each other's submissions. The public has proffered 255 pieces of prior art and 46 research suggestions. They have posted close to 500 discussion comments and put 232 tags on patent applications.[110]

The USPTO made good use of the submissions from the Peer to Patent pilot. Thirteen of the forty-six patent applications were rejected by the USPTO using Peer to Patent submission. In two cases, the public was able to guide the examiner to the references needed to make a decision. Also, "When one compares information submitted by the inventor with the application to the public submissions, the USPTO is more than twice as likely to use a Peer to Patent submission."[111]

Admittedly, the number of submissions to Peer to Patent is modest compared with Wikipedia. Out of one million registered users, 100,000 have contributed ten or more Wikipedia entries.[112] But Peer to Patent research deals with highly technical and scientific literature. The significance of the Peer to Patent project is not that it has garnered a wide user base, but that it is sufficient proof that it pays to include the relevant public/stakeholders in the discussion, even for highly scientific/technical issues.

It is also important to note that successful online participation is not only dependent on well designed applications. The experience of OECD countries shows that the three key factors for consideration when seeking to use ICTs for online citizen engagement are:

- **Timing** — At what phase in the policy-making cycle is ICT enabled participation made open?

- **Tailoring** — Are the right tools used to address the right issues and the right bodies?, and
- **Integration** — Are traditional and new channels for citizen engagement in policy-making linked?[113]

6. CREATING A REGIONAL IDENTITY

In its preamble, the ASEAN Charter declared the commitment of the peoples of Southeast Asia to intensify:

> community building through enhanced regional cooperation and integration, in particular by establishing an ASEAN Community which is comprised of the ASEAN Security Community, the ASEAN Economic Community and the ASEAN Socio-Cultural Community.

In 2009, in its summit in Cha-am, ASEAN adopted a roadmap for an ASEAN Socio-Cultural Community (ASCC). ASEAN's 2010 theme was:

> Towards an ASEAN Community: From Vision to Action, which includes steps to accelerate implementation of three blueprints, enforcement of the charter by developing a legal framework and fostering ties with countries and groups outside the bloc.[114]

Despite its success in political security and economic community building, ASEAN has yet to gain traction in socio-cultural community building efforts — the crucial third leg of the ASEAN community. It has been observed that:

> While communities of practice have been built by bringing together experts and practitioners in the many areas of ASEAN

cooperation, this has not yet been translated or expanded to generating a sense of community among the citizens of ASEAN member states.[115]

A recent meeting of ASEAN observers identified the following challenges in designing the ASEAN Socio-Community Blueprint:

- Unlike the AEC, the ASCC lacks concrete drivers, as the stakeholder are far more diverse and complex;
- While the AEC Blueprint has specific targets and timelines, it will be difficult for the ASCC Blueprint to follow the same format. As many of ASCC goals are driven by national agendas, implementation of initiatives has to be done both at national and regional levels and hence may require two types of score cards;
- ASEAN member countries may not have a common understanding of social issues, such as values, ethics, and social justice. Furthermore, unlike the European Union, ASEAN does not have the institutional capacity to translate these social issues into legislation;
- Unlike the AEC, the ASCC does not enjoy the same level of support or enthusiasm from ASEAN's dialogue partners; and
- The ASCC does not have a champion figure. Although the lead ASEAN body for the ASCC is the ASEAN Standing Committee (ASC), it is more preoccupied with political issues.[116]

The creation of the ASEAN Community Councils would, it is hoped, solve the absence of a champion figure. But there are other issues that still need to be thought through, one of which is the role of ICT in developing a regional identity and community.

Acharya argues that Southeast Asia is an "invented region".[117] It was invented by outside powers, intellectuals, and its own leaders as a "distinctive region in relation to China and India".

Acharya also forwards a strong point that identity is important to regional organizations: "Identity is what makes regional organizations constitutive, as opposed to merely regulative—communities."[118] Like the region, identity is also socially constructed:

> identity need not be cultural identity, or something rooted in primordial loyalties or immutable factors such as geography. Identity can be socially constructed even within groups whose members are culturally and politically disparate.[119]

How are regions and identities constructed? It is done by enabling people to imagine that they are part of a new community. As Acharya states, "regions are *imagined communities* just like nation-states". (italics supplied)[120]

Acharya's overt reference to Benedict Anderson's influential book on nationalism — *Imagined Communities: Reflections on the Origin and Spread of Nationalism*[121] — provides a perspective from which we can understand the role that technology, such as ICT, could play in regional community building.

For Anderson, "print capitalism" was one of the factors that gave rise to the nation state. According to Anderson:

> What, in a positive sense, made the new communities imaginable was a half-fortuitous, but explosive, interaction between a system of production and productive relations (capitalism), a technology of communication (print), and the fatality of human linguistic diversity.[122]

When booksellers started printing in the vernacular (and not Latin) in order to cater to a new market, new identities were imagined and were, consequently, born.

If "print capitalism" helped constitute the first nation state of Europe, can today's Internet, the so-called Web 2.0, help create a regional identity in Southeast Asia?

Web 2.0 applications, such as wikis, blogs, and virtual worlds, are powerful tools that can be harnessed for building a community. Writing in 2006 about the implications of the rise of Web 2.0 sites such as YouTube, Facebook (FB) and Twitter, Steven Levy and Brad Stone concluded that:

> Less than a decade ago, when we were first getting used to the idea of an Internet, people described the act of going online as venturing into some foreign realm called cyberspace. But that metaphor no longer applies. MySpace, Flickr and all the other newcomers aren't places to go, but things to do, ways to express yourself, means to connect with others and extend your own horizons. Cyberspace was somewhere else. The Web is where we live.[123]

Below are some statistics from the "future buzz" to give readers an idea of the pervasiveness of Web 2.0:

- There are 10,000,000 total articles in Wikipedia in 260 languages. (Only 2,695,205 articles on Wikipedia are in English.) These articles were contributed by 75,000 individuals. An estimated 684,000,000 visited Wikipedia in 2009.
- In March 2008, there is an estimated 70,000,000 videos on YouTube (which would take an individual 412.3 years to view). These videos were uploaded by 200,000 video publishers

whose average age is 26.57. About 100,000,000 YouTube videos are viewed per day.

- There are an estimated 200,000,000 active Facebook users, half of whom (100,000,000) log on to Facebook at least once each day. These users are from 170 countries/territories using thirty-five different languages.
- There are 133,000,000 blogs that have been indexed by Technorati since 2002. Daily, there are 900,000 blog posts. These blogs are in eighty-one languages and read by around 346,000,000 individuals globally.[124]

According to the research firm, IDC, there are "110 million Internet users across Southeast Asia" in 2009.[125] IDC's Net Index 2010 provides more details:

1. Mobile access to the Internet is gaining momentum, with more new to Internet users jumping directly into this space;
2. Online gaming is developing into a dominant activity within the entertainment category in markets such as Vietnam;
3. Social media is changing the rules of online engagement and is ushering in a fundamental shift in the way people interact with one another and media.
4. The popularity of online news has reached new heights for a number of reasons, especially in Vietnam.[126]

Among the FB users in Asia are the ASEAN leaders. Nine out of ten ASEAN heads of state or government have FB pages.[127] Indonesian President Susilo Bambang Yudhoyono, Malaysian Prime Minister Najib bin Tun Abdul Razak, and Thai Prime Minister Abhisit Vejjajiva's FB pages have over 500,000 "likes".

Given the Web's increasing pervasiveness, should not ASEAN consider using Web 2.0 to help its citizens "imagine" an ASEAN community?

The *blog* is an inexpensive way for ASEAN to reach out to one's constituency directly.

In fact, the importance of a blog is something that Singapore's foreign minister understands. Then Foreign Minister George Yeo blogs at *BeyondSG*, which is described as "a blog about Singapore and our social and business connectivity with the world".[128]

While then Minister Yeo's blogs comprise his speeches and official statements, his more interesting posts are reflections on these trips — "Visit to Xiamen", posted on 22 December 2009; "Return to Cuba", 5 December 2009; "Visit to Manila", 20 November 2009; comments on movies he's seen — "Avatar", posted on 27 December 2009 — and people he has met — Jacques Attali, posted on 5 December 2009, and Sartaj Aziz, posted 5 November 2009. As to be expected of an elected official, he discusses constituency issues in his blog e.g. "Lift upgrading", posted on 7 November 2009.

The power of a blog to make an organization and its leaders accessible to the public is evident in Yeo's "Aborted Summit in Pattaya" blog.[129] In this blog — including photos — one gets a unique front row seat into one of the most dramatic behind-the-scene events in ASEAN's history — the evacuation of the ASEAN leaders in the aborted April 2009 ASEAN Summit.

In summary, Yeo's blog provides him with an unmediated platform to reach his constituency. Furthermore, the informality and immediacy of his blogs (as well as his Facebook account, and YouTube uploads) project the image of an accessible and hardworking leader.

If adopted, a blog would provide the ASEAN secretary general with a platform to address the citizens of ASEAN member-states directly. Instead of relying on the mass media to pick up a press release prepared by the Secretariat, the secretary general can go straight to the ASEAN public to share the latest position of ASEAN on urgent issues.

Aside from blogs, a *wiki* is an inexpensive and unutilized tool for ASEAN community building.

In Canada, a national newspaper — *Globe and Mail* — and a non-profit public policy centre — The Dominion Institute — launched a site called the Public Policy Wiki in 2009. The goal is to improve the dialogue between citizens and governments using social networking (Web 2.0) tools. The first topic for discussion — the Canadian economy and the federal budget — elicited "100,000 page views and more than 800 readers signed up and voted, commented and edited — and created over 30 detailed briefing notes".[130] The two most popular notes were sent to the finance minister. One of those who helped launch the project assessed the project's first effort:

> Getting 800 people to register may not seem like a lot (although 100,000 page views in just two weeks is pretty respectable), and there's no question that we were hardly bombarded with traffic to the site. But I think that is a result of a number of factors — including the fact that not everyone feels passionate enough about the economy and the budget to put together a detailed policy proposal. That said, I think the issue we chose and the serious way in which we framed the project helped to encourage thoughtful responses, and in fact the vast majority of the contributions we've gotten so far have been intelligent, well thought-out and (for the most part) well intentioned.[131]

A similar effort in the United States is President Obama's Citizen's Briefing Book. This effort provided a venue for Americans to post, comment, and vote on ideas for the then newly elected president.[132] The ideas with the highest votes will be presented to the U.S. president. The result of the initiative is described below:

> Between the election and the inauguration, the Obama transition web site — Change.gov — accepted over 44,000 ideas from over 125,000 people on what President Obama should do to improve the nation and the federal government. As the peoples' questions were posted at Change.gov, all Americans were allowed to vote for the ideas they most liked. Based on over 1.4 million votes, with the most popular ideas accumulating tens of thousands of votes each.[133] (*sic*) The Citizens Briefing Book presents some of the top vote-getting ideas, broken into groups by issue area.

The Canadian and U.S. examples show that citizens are more than willing to share their views on governance issues when asked and provided with a mechanism to do so easily.

ASEAN should also seriously consider addressing youth in the language and channel that they prefer. In particular, ASEAN should learn how to harness social networking sites and an online or web-based community for community and identity building purposes.

Social network sites (SNS) are

web-based services that allow individuals to:
1) construct a public or semi-public profile within a bounded system;
2) articulate a list of other users with whom they share a connection; and
3) view and traverse their list of connections and those made by others within the system.[134]

While most are familiar with the English-language social network sites such as Facebook and Twitter, it is important to note that there are also huge non-English SNS such as CyWorld (Korean), Mixi (Japanese), and QQ (Chinese). In fact, the "usage of Chinese social media is some of the most intense in the world".[135] Interestingly, while there are differences between the Chinese- and English-language SNS, the "variation is not all due to censorship ... local variations of Internet usage are driven by language, culture, levels of economic development, and the underlying digital ecosystem".[136]

A recent study of Internet users in the Asia-Pacific region reveals that the youth in ASEAN are significant users of social network sites (see Table 1).[137] Ninety per cent of young Filipinos, 89 per cent of Indonesians, 85 per cent of Malaysians, and 84 per cent of Singaporeans use social network sites.

This study also revealed that Facebook was the most popular social network in the Philippines (84.5 per cent market share), Indonesia (84.9 per cent), Malaysia (77.5 per cent), Singapore (72.1 per cent), and Vietnam (18.4 per cent).

It is also important to note that this study probably underestimates reality as it excludes Internet access from public computers such as in Internet cafes, or access from mobile phones or personal digital assistants (PDAs) — two of the most popular modes of accessing the Internet in Southeast Asia.

SNS can serve as an inexpensive and reliable platform for greater people-to-people exchange in ASEAN. For instance, women entrepreneurs in ASEAN can set up a page on a SNS where they can update themselves on events and other activities. Lawyers, accountants, and other professionals could also benefit

TABLE 1
Social Networking Reach and Engagement in ASEAN[138]

Total Internet Audience*, Age 15+ — Home & Work Locations

Social Networking

	% Reach	Rank (based on the 12 countries in the study	Ave. minutes per Visitor	Ave. visits per Visitor
Asia Pacific	**50.8**		**148.9**	**15.1**
Philippines	90.3	1	332.2	26.3
Indonesia	88.6	3	324.4	22.6
Malaysia	84.7	4	226.0	22.3
Singapore	83.7	5	220.9	22.1
Vietnam	46.1	11	49.5	7.2

Source: "Social Networking Habits Vary Considerably Across Asia-Pacific Markets", comScore, 7 April 2010 <http://comscore.com/Press_Events/Press_Releases/2010/4/Social_Networking_Across_Asia-Pacific_Markets> (accessed 20 April 2011).

from ASEAN-focused SNS pages. But the main target for social networking should be ASEAN youth.

Various studies have pointed to the importance of SNS to the youth sector of society. SNS has become an alternative source of information and entertainment for today's youth. The so-called Net generation does not get their news from newspapers, but from the Internet, particularly SNS. Moreover, they prefer watching YouTube to television. Facebook and QQ are supplanting email as communications media.

ASEAN needs to harness the power of SNS for its own community building efforts. Otherwise, it may become irrelevant if it is unknown to its youth. ASEAN can build on the Secretariat's ASEAN VOICES initiative or host an island in Second Life where information of ASEAN activities can be made readily available to its youth.

Online or web-based communities are "people who come together for a particular purpose, and who are guided by policies (including norms and rules) and supported by software".[139] Studies have shown that online communities have the capacity to increase social ties and interaction. A wide array of topics is covered in online communities. In these communities:

> People engage in social interaction with others, establishing their identity in their profiles, responding to the identity of others, complementing social relationships, information seeking, and completing tasks.[140]

Online community members view the technology they use as "a flexible form of their own expression".[141]

The most interesting online communities are "virtual worlds". A virtual world:

> ... is a genre of online community that often takes the form of a computer-based simulated environment, through which users can interact with one another and use and create objects. Virtual worlds are intended for their users to inhabit and interact, and the term today has become synonymous with interactive 3D virtual environments, where the users take the form of avatars visible to others graphically. These avatars are usually depicted as textual, two-dimensional, or three-dimensional graphical representations, although other forms are possible (auditory and touch sensations, for example).[142]

Among the most documented virtual world is Second Life, which was launched on 23 June 2003 and is accessible via the Internet.[143] In Second Life, users interact with one another through avatars. Users socialize, participate in individual and group activities, and create and trade virtual property and services with one another, or travel throughout their created world.

Second Life is not just a game. The currency used in Second Life is convertible into U.S. dollars. At least 300 universities from around the world teach courses or conduct research in Second Life, which also has a "Diplomacy Island" where visitors are able to talk face-to-face with a computer generated ambassador about visas, trade, and other issues. Maldives was the first country to open an embassy in Second Life. Sweden and Estonia shortly followed suit. Other countries with Second Life embassies are Colombia, Serbia, Macedonia, and the Philippines. However, these embassies are used to promote their respective country's image and culture, rather than provide any real or virtual services.

A specific action that ASEAN could initiate is to establish an island in Second Life where visitors can learn about its current initiatives and future goals. *SL-ASEAN* can be a preview of the much desired "caring community with a heart".

If an ASEAN island in Second Life is too radical a proposition, perhaps ASEAN can follow IBM's lead in using Second Life as a platform for an international conference.

In late 2008, IBM's Academy of Technology held a Virtual World Conference hosted in a secure Second Life environment.[144] The conference space was specially designed by IBM for "keynotes, breakout sessions, a simulated Green Data Centre, a library, and various areas for community gathering". The three-day Virtual World Conference attracted over 200 members worldwide who attended three keynotes and thirty-seven breakout sessions. IBM estimates the return of investment for the Virtual World Conference at roughly $320,000. While the company invested $80,000 for the development of the virtual conference centre, they saved over $250,000 in travel and venue costs and an additional $150,000 in "productivity gains (since participants were already at their computers and could dive back to work immediately)". According to IBM Academy of Technology President Joanne Martin:

> Second Life provided an opportunity for us to have a positive social and technical exchange, addressing most of our collaboration objectives. And, we delivered the experience at about one fifth the costs and without a single case of jet lag.

But could ministers use avatars to discuss security or economic issues? Maybe not. But the younger diplomats and civil servants of ASEAN would probably be more open to the idea — if it means less travel and more work done. A more acceptable proposition and "cool" idea would probably be an ASEAN youth meeting held in Second Life.

The European Parliament is cognizant of the value and function of role playing games and social networking to reach youth. It developed Citizalia — "a role playing game and social networking forum wrapped in a virtual 3D world that captures the essence of the European Parliament".[145] Citizalia is described as:

> ... a means of understanding how the EU's democratically elected Parliament works. By participating in debates on issues the European Parliament is currently discussing you will gain insight and expertise into how democracy works in the EU. It is an opportunity to hear how other fellow citizens feel about current issues and about the role of the European Parliament. It is also a platform for debate and discussion of the issues which have been, are or can be addressed by the European Parliament.[146]

Virtual worlds can be powerful tools to help develop new identities. If the ASEAN community is to become reality, it must first be imagined. More and more people are dreaming of an ASEAN community, but by most accounts, the numbers are not nearly enough. It should be time for ASEAN to use new technologies to address the next generation (for whom these technologies are not really new — just current). As Tim Guest explains, on the subject of virtual worlds:

> Throughout millennia, mankind dreamed alone. Now, suddenly, in our virtual worlds, we were able to share and inhabit each other's dreams.[147]

7. ICT AND NETWORK MANAGEMENT

ASEAN faces the challenge of how to evolve into an organization that is capable of effectively and efficiently realizing its goal of regional integration. This study argues that it can be done if ASEAN exploits its character as a network organization. In particular, this study points to the important role ICT plays in responding to the challenge of regional integration and community building. So far, this has been substantiated through a discussion of the role of ICT in three areas:

1) horizontal and vertical coordination;
2) the inclusion of non-government organizations in policy making; and
3) the creation of an ASEAN identity and community.

This section will be devoted to a discussion of ICT and network management.

Scholars have recognized the importance of network management for intergovernmental organizations. Some argue that "inter-organisational networks centralised around a primary coordinating agency — a network administrative organisation — produce better outcomes".[148] Jordan and Schout, who prefer the term, "network manager", recognize the importance of this agency "to the stability of the network and to the handling of issues with high transaction costs".[149]

Jordan and Schout posit that the "network administrative organisation" can act as a "secretary" whose role is to achieve "greater efficiency in decision making" by "gathering and distributing information as part of the benchmarking and peer review exercises".[150] Or, the organization can serve as network manager — one who "seeks actively to manage coordination in the network".[151] The network manager arranges and facilitates the "interaction process within networks in such a way that problems of under or non-representation are properly addressed and interests are articulated and dealt with in an open, transparent and balanced manner".[152] Network management involves three activities:

1) intervening in existing patterns/restructuring of network relations;
2) furthering conditions for cooperation; consensus building; and
3) joint problem solving.[153]

Joaquin Herranz identifies four archetypes of the network manager: reactive facilitator; contingent coordinator; active coordinator; and hierarchical-based network administrator.[154]

In reactive facilitation, "effective network governance is conceived less as managerial intervention and more as a relatively passive stewardly facilitation of collective solutions".[155] In contingent coordination, "managers may exert some coordinating influence on networks, but the scope of managerial behaviour is limited and contingent upon network interests, resources, and opportunities".[156] In active coordination, the network manager acts as a "network 'integrator' who establishes communication channels; coordinates activities between network participants so that they share knowledge, aligns values and incentives, overcomes cultural differences; and builds trusting relationships".[157] Here "network leadership requires honest brokering, stabilising coordination, and structural management".[158] In the hierarchical-based network administration, the role of a network manager is the same as the department head.[159] The premise is that networks and hierarchies require the same managerial skills.

If one uses Herranz' passive-to-active continuum of the network manager, one could see that the ASEAN Secretariat was originally conceived as a reactive facilitator. However, it has evolved into the contingent coordinator role. While ASEAN history and ideology may prevent the Secretariat from evolving into a hierarchical-based network administrator, the question that ASEAN must face today is this: will it allow its Secretariat to become an active coordinator?

ICT helps in all the various aspects of network management and aids the network manager (in whatever role) to discharge his/her duties effectively. An ICT infrastructure/ecosystem that facilitates sharing and exchanging of information can help a network manager build relationships and interactions that result in achieving the network purpose.[160] Moreover, ICT can "reduce complexity and uncertainty ... (resulting) in stronger linkages among the actors within and outside of the network".[161]

The critical role of technology in network management demands that the network has an individual who will take the lead in development and implementation. The job title of this individual is chief information officer or CIO.

Traditionally, the CIO's main responsibility is to manage an organization's IT resource and operations.[162] This role has evolved to include contributing "to the effective use of information and knowledge in (the organization's) units by developing appropriate information, people and IT capabilities". Aside from engineering ICT, the CIO should also help "the organisation adapt as technology speeds up the exchange of more information, integrated disjointed work processes and digitised services".[163] The CIO ensures the alignment between technology and the organization's strategies.

In the network organization, the CIO's principal task is to supply the technical infrastructure for network governance.[164] He/she is expected to act as the network integrator with the following responsibilities:

- *Coordinating Activities* — Most agencies still interact with partners through manual processes, creating a host of inefficiencies ranging from slow responsiveness and poor reliability to uncoordinated service delivery. The CIO should take the lead in addressing these problems;
- *Synchronising Response* — The CIO should lead in the development and deployment of an electronic coordination mechanism that allows disparate groups to share information in real time and synchronize their responses;
- *Providing A Single Client View* — The CIO should create a system that allows network partners to share relevant "customer" information to coordinate activities;
- *Sharing Knowledge* — The CIO must lead in dismantling hierarchical systems that instil data control and create the

technical infrastructure that let people share information and knowledge across geographic and organizational boundaries; and

- *Measuring Performance* — The CIO should exploit technology to measure and track performance within a complex network.

The most urgent tasks for the ASEAN CIO are laying the foundation for electronic interaction between ASEAN processes and partners (coordinating activities), and deploying the electronic coordination mechanism that allows disparate groups to share information in real time (synchronizing response). With these goals, the ASEAN CIO should focus on interoperability.

Interoperability is "the ability to exchange and use information" (usually in a large heterogeneous network made up of several local area networks).[165] It also means:

> guaranteeing a media-consistent flow of information between citizens, business, the government and its partners and selecting only those specifications that are relevant to systems' interconnectivity, data integration, e-services access and content.[166]

In the public sector, there are at least two ways to secure interoperability. The first is through a Government Interoperability Framework or GIF. A GIF "is a set of standards and guidelines that a government uses to specify the preferred way that its agencies, citizens and partners interact with each other".[167] The GIF includes all the technical specifications that would allow information reuse across the government and within government interactions with all stakeholders. Australia, Brazil, Denmark, Germany, Malaysia, New Zealand, and the United Kingdom are among the countries which have Government Interoperability Frameworks.[168] The European Union has a European Interoperability Framework (EIF) that was adopted in 2004 and revised in 2010.[169] The EIF aims:

- To promote and support the delivery of European public services by fostering cross-border and cross-sectoral interoperability;
- To guide public administrations in their work to provide European public services to businesses and citizens; and
- To complement and tie together the various National Interoperability Frameworks (NIFs) at European level.[170]

For the EIF, "interoperability is multilateral in nature and is best understood as a shared *value* of a community".[171]

The other approach to interoperability is through Enterprise Architecture (EA) — "a strategic planning framework that relates and aligns ICT with the business functions that it supports".[172] Also called National Enterprise Architecture or Federal Enterprise Architecture and even Governance Enterprise Architecture, the EA is a common framework that ensures:

1) general coherence between public-sector IT systems;
2) the systems are optimized in terms of local needs; and
3) quality improvement, resource optimization, and cost reduction.[173]

Above all, the enterprise architecture ensures "systems interoperability even if administrative branch-specific information systems are maintained".[174]

In 1999, the CIO Council of the U.S. Government developed the Federal Enterprise Architecture (FEA) framework. This framework serves:

> as a reference point to facilitate the efficient and effective coordination of common business processes, information flows, systems, and investments among Federal Agencies and other Governmental entities.[175]

Though this framework, "Government business processes and systems will operate seamlessly in an enterprise architecture that provides models and standards that identify and define the information services used throughout the Government."

A recent development that the ASEAN CIO needs to consider is the emergence of "cloud computing". *The Economist* pointed out that the advantage of cloud computing is that it "allows firms in developing countries to leapfrog traditional information technology (IT) and benefit from advanced computing services without having to build expensive infrastructure".[176] Cloud computing also makes for a mobile workforce.

The U.S. Government has already embraced cloud computing, which it defines as:

> a model for enabling convenient, on-demand network access to a shared pool of configurable computing resources (e.g., networks, servers, storage, applications, and services) that can be rapidly provisioned and released with minimal management effort or service provider interaction.[177]

For the U.S. Government, the advantages of cloud computing are:

- **Significant Cost Reduction**: Cloud computing is available at a fraction of the cost of traditional IT services, eliminating upfront capital expenditure and dramatically reducing the administrative burden on IT resources.
- **Increased Flexibility**: Cloud computing provides on-demand computing across technologies, business solutions, and large ecosystems of providers, reducing time to implement new solutions from months to days.
- **Access anywhere**: One is no longer tethered to a single computer or network. One can change computers or move to portable devices, and one's existing applications and documents follow one through the cloud.
- **Elastic scalability and pay-as-you-go**: This allows one to add and subtract capacity depending on what is necessary. Pay for only what is used.
- **Easy to implement**: There is no need to purchase hardware, software licences, or implementation services.
- **Service quality**: Cloud computing service providers offer reliable services, large storage and computing capacity, and 24/7 service and up-time.
- **Delegate non-critical applications**: Cloud computing provides a way to outsource non-critical applications to service providers, allowing agency IT resources to focus on business-critical applications.
- **Always the latest software**: This eliminates the problem of choosing between obsolete software and high upgrade costs. When the applications are web-based, updates are automatic and are available the next time you log on to the cloud.
- **Sharing documents and group collaboration**: Cloud computing lets users access all applications and documents from anywhere in the world, freeing users from the confines of

the desktop and facilitating group collaboration on documents and projects.

One way that the U.S. Government encourages the uptake of cloud computing is through Apps.Gov, an Internet portal built "to allow government organisations quick and painless access to purchasing cloud services for their operations".[178] Apps.Gov offers government agencies four types of solutions:

- *Business Apps* — cloud software solutions such as analytical, business processes, CRM, tracking and monitoring tools, and business intelligence.
- *Cloud IT Services* — cloud hosted storage, web hosting, and virtual machines.
- *Productivity Apps* — cloud hosted applications for civil servants who need software to perform daily tasks such as word processing and spreadsheets. This also includes brainstorming, collaboration, document management, and project management.
- *Social media Apps* — various online technology tools that enable people to communicate easily and share information; includes text, audio, video, images, podcasts, and other multimedia communications.[179]

Cloud computing is no longer a futuristic scenario. Analysts forecast that in 2011 there will be a "massive wave of adoption, innovations and transformation as the cloud crosses the chasm from the early adopters to larger, more pragmatic organisations".[180]

The above is merely indicative of the issues that the ASEAN CIO faces. There are other technical challenges that the CIO and, indeed, the leadership of network organizations such as ASEAN face. Building databases that can provide integrated support for

ASEAN officials and "citizens" alike, and ensuring online security and privacy, are only some of these. The other challenges for ASEAN are more fundamental and not only ICT related. Aside from creating an ICT-competent workforce in the Secretariat, the most critical ICT challenge for ASEAN's leadership is establishing new values, attitudes, and behaviours concerning information sharing and collaborative decision making in the organization.

8. ICT AND ASEAN'S CONTINUING RELEVANCE

ASEAN's champions and critics alike believe that in order to remain relevant, the organization must change. For Acharya, "ASEAN should seize the moment or fade away as a new regional and global leadership takes over."[181] Acharya believes that ASEAN's ability to implement the ASEAN community blueprint faithfully and its capacity to deal with new challenges will determine its success or failure. Others, such as Thitinan Pongsudhirak, believe that ASEAN's future lies in its ability to transform from a "top-down intergovernmental organisation driven by policymaking elites and bureaucrats" to a "bottom-up process via the third-track of people-to-people enmeshment" as envisioned in its new charter.[182]

This study shares the view that ASEAN must be able to respond decisively to changes in the global order (marked by globalization) and changing constellations of actors (emergence of global, non-state actors) to continue to be relevant.

The main argument of this study is that ASEAN should harness ICT to enhance its network form of organization and thereby become more effective and remain relevant.

Following Castells, this study believes that networks are "the most flexible, and adaptable forms of organisation, able to evolve with their environment and with the evolution of the nodes that compose the network".[183] Historically, networks are weak in coordination, particularly, the management of complexity. But this historic weakness has been superseded by the use of ICT by networks. Castells believes that ICT not only allows networks to keep their flexibility and adaptability, but ICT also allows:

> for co-ordination and management of complexity, in an interactive system which features feedback effects, and communication patterns from anywhere to everywhere within the networks. It follows an unprecedented combination of flexibility and task implementation, of co-ordinated decision making, and de-centralised execution, which provide a superior social morphology for all human action.[184]

ICT could become ASEAN's most useful tool in achieving its goal of "One Vision, One Identity, One Community". Like any tool, ICT can be used inappropriately. But if wielded effectively, ICT can help ASEAN achieve its goal of regional integration and community building by enabling the following:

1) effective policy coordination;
2) enhanced participation of stakeholders;
3) intensified community building efforts by helping to create a regional identity, particularly among its youth; and
4) effective network management.

The ASEAN Charter created new regional bodies to enhance the organization's policy-making and coordination efforts. However, creating more committees does not necessarily lead to greater policy coordination. This may only bring about more complex processes and increased transaction costs. It was argued that more intensive use of ICT by these new ASEAN bodies, including the ASEAN Secretariat, could facilitate better policy coordination at reduced costs.

At present, ICT use in the ASEAN organization is limited to office productivity tools (email, word processing, and spreadsheets) and a web browser. ASEAN should examine new *software-based conferencing solutions* for audio- and web-conferencing to complement existing face-to-face ASEAN meetings. Private sector companies that use these capabilities have achieved increased efficiencies, total cost savings, and environmental footprint reductions. Similar outcomes can be expected if ASEAN begins using these conferencing tools.

ICT can help ASEAN achieve its goal of becoming an inclusive institution of governance. ASEAN stakeholders represent a resource — expertise on wide ranging issues that is currently not maximized in ASEAN policy-making. e*Participation* — the use of ICT for enabling appropriate stakeholder participation in decision making — could change this state of affairs.

ICT can help promote/create a regional identity by enabling the citizens of member states to "imagine" an ASEAN community through SNS. SNS can serve as inexpensive and reliable platforms for greater people-to-people exchange in ASEAN. For instance, women entrepreneurs in ASEAN can set up a page on a SNS where they can update themselves on events and other activities. Lawyers, accountants and other professionals could also benefit from ASEAN-focused SNS pages. But the main target for social networking should be the ASEAN youth.

Specifically, ASEAN should consider adopting the following:

- An appropriate, robust, cost efficient, common *conferencing* and/or *collaborative software* to be used by various ASEAN bodies/meetings;
- e*Participation applications/initiatives* to deepen engagement with ASEAN stakeholders;

- Policy *blogs* and *wikis* as part of ASEAN's knowledge management and information dissemination strategy; and,
- Implement a Social Networking Strategy aimed at ASEAN youth. This strategy should leverage the power and appeal of social networking tools to develop an ASEAN identity among its youth.

The full promise of these initiatives can only be realized if these recommendations are embedded within an overall, integrated ICT framework to be implemented by competent individuals. Hence, from a network management perspective, there is a need to appoint an ASEAN CIO who will provide leadership for developing and implementing ASEAN eGovernance applications.

Finally, the successful use of ICT by ASEAN is not dependent on and limited to correct technology choices or the hiring of a competent CIO. ASEAN officials at all levels, particularly the highest levels of the organization, must fully recognize ICT as a tool for enhanced governance, effective policy-making, and intensified community building efforts in Southeast Asia.

Notes

1. Media release, "ASEAN Leaders Sign ASEAN Charter", Singapore, 20 November 2007 <http://www.aseansec.org/21085.htm> (accessed 22 December 2009).

2. "ASEAN Economic Blueprint Aims to Harmonise Business Rules", at <http://www.channelnewsasia.com/stories/singaporelocalnews/view/310873/1/.html> (accessed 22 December 2009).

3. "ASEAN Political-Security Community Blueprint", at <http://www.aseansec.org/5187-18.pdf> (accessed 6 January 2010).

4. *The ASEAN Community: Unblocking the Roadblocks*, ASEAN Studies Center Report No. 1 (Singapore: Institute of Southeast Asian Studies, 2008), p. 10.

5. Amitav Acharya and Alastair Iain Johnston, "Comparing Regional Institutions: An Introduction", in *Crafting Cooperation: Regional International Institutions in Comparative Perspective*, edited by Amitav Acharya and Alastair Iain Johnston (Cambridge: Cambridge University Press, 2007), p. 5.

6. Barbara Koremenos, Charles Lipson, and Duncan Snidal, "The Rational Design of International Institutions", *International Organization* 55, no. 4 (Autumn 2001): 762.

7. Acharya and Johnston, "Comparing Regional Institutions", pp. 15–16.

8. Erik-Hans Klijn and Joop F. M. Koppenjan, "Institutional Design: Changing Institutional Features of Networks", *Public Management Review*, p. 148.

9. Ibid.

10. Ibid., p. 149.

11. Ibid., p. 155.

12. Laurence J. O'Toole Jr., "Treating Networks Seriously: Practical and Research Based Agendas in Public Administration", at <http://www.questia.com/PM.qst?a=o&se=gglsc&d=5000414319> (accessed 20 April 2011).

13. Manuel Castells, "Materials for an Exploratory Theory of the

Network Society", *British Journal of Sociology* 51, no. 1 (January/ March 2000): 15.

14. Mark Considine, *Making Public Policy: Institutions, Actors, Strategy* (Cambridge: Polity, 2005), p. 129.

15. Charlotte Streck, "Governments and Policy Networks: Chances, Risks, and a Missing Strategy", p. 4 at <http://www.climatefocus. com/downloads/publications/gov_policy_networks_cstreck.pdf> (accessed 20 April 2011).

16. Candace Jones, William Hesterly and Stephen Borgatti, "A General Theory of Network Governance: Exchange Conditions and Social Mechanism", *Academy of Management Review* 22, no. 4 (1997): 914.

17. Chris Ansell, "The Networked Polity: Regional Development in Western Europe", in *Governance: An International Journal of Policy and Administration* 13, no. 3 (July 2000): 305, at <http:// www.unc.edu/depts/europe/conferences/mlg/papers/ansell_c.pdf> (accessed 27 December 2009).

18. Ibid., p. 308.

19. Andrew Jordan and Adriaan Schout, *The Coordination of the European Union: Exploring the Capacities of Networked Governance* (Oxford: Oxford University Press, 2006), pp. 6–7. It is also important to underscore that Jordan and Schout did not specifically consider the role of ICT in the coordination of networked governance.

20. Castells, "Materials for an Exploratory Theory of the Network Society", p. 15.

21. W. Jansen, GCA Steenbakkers, and H. Jagers, "Coordination and Use of ICT in Virtual Organizations", PrimaVera Working Paper 98-05, at <http://primavera.feb.uva.nl/PDFdocs/98-05.pdf> (accessed 1 April 2011).

22. Philip G. Cerny, "Globalization, Governance, and Complexity", in *Globalization and Governance*, edited by Aseem Prakash and Jeffrey A. Hart (New York: Routledge, 1999), p. 188.

23. <http://en.wiktionary.org/wiki/eGovernment> (accessed 20 April 2011).

24. Richard Heeks, "Understanding e-Governance for Development", i-Government Working Paper Series No. 11, at <http://www.sed.manchester.ac.uk/idpm/research/publications/wp/igovernment/igov_wp11.htm> (accessed 20 April 2011).

25. <http://www.finance.gov.au/e-government/strategy-and-governance/gov2.html> (accessed 27 December 2009).

26. Don Tapscott and Anthonny D. Williams, *Macrowikinomics: Rebooting Business and the World* (London: Atlantic Books, 2010), p. 264.

27. Ibid.

28. <http://en.wiktionary.org/wiki/Web_2.0> (accessed 27 December 2009).

29. <http://en.wikipedia.org/wiki/Blog>.

30. Julianne Mahler, "Agency-Related Blogs as Forums for Policy Networks", paper presented at the annual meeting of the APSA 2008 Annual Meeting, Hynes Convention Center, Boston, Massachusetts, 28 August 2008, p. 3, at <http://www.allacademic.com/meta/p280172_index.html> (accessed 27 December 2009).

31. Ibid., p. 4.

32. <www.govst.edu/elearning/default.aspx> (accessed 27 December 2009).

33. Mathew Ingram, "The Policy Wiki: A Social Experiment", at <http://www.niemanlab.org/2009/01/the-policy-wiki-a-social-experiment/#more-1381> (accessed 20 April 2011).

34. Jennifer Gonzalez-Reinhart, "Wiki and the Wiki Way: Beyond a Knowledge Management Solution", Information Systems Research Center, February 2005, p. 10, at <http://www.uhisrc.com/FTB/Wiki/wiki_way_brief%5B1%5D-Jennifer%2005.pdf> (accessed 20 April 2011).

35. "Diplopedia — Wikipedia-style Diplomacy — A Success at US Department of State", *Science Daily: News & Articles in Science, Health, Environment & Technology*, 20 May 2010, <http://www.sciencedaily.com/releases/2010/05/100520112345.htm> (accessed 20 April 2011).

36. <http://www.bakerinstitute.org/publications/TSPP-pub-BronkSmith DiplopediaDraft-051810.pdf> (accessed 21 November 2010).
37. <http://www.state.gov/m/irm/ediplomacy/115847.htm> (accessed 21 November 2010).
38. See Government 2.0 Best Practices Wiki, at <http://government20 bestpractices.pbworks.com/w/page/10044435/FrontPage> (accessed 21 November 2010).
39. <http://www.futuremelbourne.com.au/wiki/view/FMPlan/ S1aAboutFutureMelbourne2020> (accessed 21 November 2010).
40. <http://amherst.localocracy.org/faq> (accessed 21 November 2010).
41. Beth Simon Noveck, *WikiGovernment: How Technology Can Make Government Better, Government Stronger and Citizens More Powerful* (Washington, D.C.: Brookings Institution Press, 2009), p. 40.
42. John Dryzek, *Deliberative Global Politics: Discourse and Democracy in a Divided World* (Cambridge: Polity Press, 2006).
43. Thus we will not specifically touch on the issues related to ICT development in the ASEAN member states — the topic covered by the ASEAN Telecommunications and IT Ministers Meeting and its predecessors.
44. Eva Sørensen and Jacob Torfing, "Network Governance and Post-liberal Democracy", *Administrative Theory & Praxis* 27, no. 2 (2005): 203.
45. Hadi Soesastro, "Lessons from ASEAN Institution", *APOFTTES. revista mexicana de estudios sobre la Cuenca del Pacifico* 2, no. 3 / Enero (Junlo 2002): 23–42.
46. Sørensen and Torfing, "Network Governance and Post-liberal Democracy", p. 209.
47. Thanat Khoman, "ASEAN Conception and Evolution", Bangkok, 1 September 1992.
48. Carolina G. Hernandez, "Institution Building through an ASEAN Charter", at <http://www.kas.de/upload/auslandshomepages/

singapore/Hernandez_AseanCharta.pdf> (accessed 20 April 2011).

49. Kripa Sridharan, *Regional Cooperation in South and Southeast Asia* (Singapore: Institute of Southeast Asian Studies, 2007), p. 56.

50. *The Founding of ASEAN*, at <http://www.aseansec.org/20024.htm> (accessed 27 December 2009).

51. Soesastro,"Lessons from ASEAN Institution".

52. Sridharan, *Regional Cooperation in South and Southeast Asia*.

53. <http://www.aseansec.org/about_ASEAN.html> (accessed 27 December 2009).

54. Volker Rittberger, "Global Governance: From 'Exclusive' Executive Multilateralism to Inclusive, Multipartite Institutions", *Tübinger Arbeitspapiere zur Internationalen Politik und Friedensforschung Nr. 52*, p. 2.

55. Ibid.

56. Hadi Soesastro, "ASEAN and Institutions for Regionalism in Asia", at <www.adb.org/AnnualMeeting/2009/seminars/hsoesastro-presentation.pdf> (accessed 20 April 2011).

57. Yuen Foong Khong and Helen E. S. Nesadurai, "Hanging Together, Institutional Design, and Cooperation in Southeast Asia: AFTA and the ARF", in *Crafting Cooperation*, p. 49.

58. "Governance in EU and ASEAN", p. 16.

59. Acharya, "ASEAN at 40".

60. Ali Alatas, "The ASEAN Charter as Legal Basis and Guiding Norm for ASEAN Cooperation", Southeast Asian Conference on Consumer Protection 2006, Bali, 5–7 November 2006.

61. Geoffrey Cockerham, "Regional Integration in Southeast Asia: Institutional Design and the ASEAN Way", paper presented at the annual meeting of the ISA's 50th Annual Convention, "Exploring the Past, Anticipating the Future", New York Marriott Marquis, New York City, U.S., 15 February 2009, at <http://www.allacademic. com/meta/p314277_index.html> (accessed 20 April 2011).

62. Tommy Koh, "ASEAN Charter at One: A Thriving Tiger Pup", *The*

Malaysian Insider, 23 December 2009, at <http://www.themalaysian insider.com/index.php/opinion/breaking-views/45888-asean-charter-at-one-a-thriving-tiger-pup–tommy-koh> (accessed 21 November 2010).

63. Klijn and Koppenjan, "Institutional Design", p. 150.

64. David A. Currie, Gerald Holtham, and Andrew Hughes Hallett, "The Theory and Practice of International Policy Coordination", in *Policies in an Interdependent World*, edited by Ralph Brynant et al. (Washington, D.C.: International Monetary Fund, 1989), p. 42, at <http://books.google.com/books?hl=en&lr=&id=Hbvmcf53VwAC&oi=fnd&pg=PA14&dq=%22Currie%22+%22International+policy+coordination%22+&ots=sviwDfNs13&sig=_dhQEif2R0i0qC0pTdC-eIv5YjM#v=onepage&q=%22Currie%22%20%22International%20pol icy%20coordination%22&f=false> (accessed 4 January 2010).

65. Ibid.

66. Antti Pelkonen, Tuula Teräväinen, and Suvi-Tuuli Waltari, "Assessing Policy Coordination Capacity: Higher Education, Science, and Technology Policies in Finland", *Science and Public Policy*, May 2008, p. 242, at <http://www.britannica.com/bps/additionalcontent/18/32507647/Assessing-policy-coordination-capacity-higher-education-science-and-technology-policies-in-Finland#> (accessed 21 November 2010).

67. S. Pushpanathan, "ASEAN Charter: One Year and Going Strong", *Jakarta Post*, 22 December 2009, at <http://www.thejakartapost.com/news/2009/12/22/asean-charter-one-year-and-going-strong.html> (accessed 20 April 2011).

68. *The Coordination of the EU*, p. 7.

69. *Horizontal Policy Coordination*, pp. 17–19.

70. Ian Peach, "Managing Complexity: The Lessons of Horizontal Policy-Making in the Provinces", Public Lecture at the Saskatchewan Institute of Public Policy, 8 June 2004, at <http://www.uregina.ca/sipp/documents/pdf/SS_Ian%20Peach_%20SpringSummer%2004.pdf> (accessed 27 December 2009), p. 27.

71. Ibid., pp. 27–29.
72. R. Kaiser and H. Prange, *The OMC in Innovation Policy*, p. 255.
73. Eric K. Clemons, Sashidhar P. Reddi, and Michael C. Row, "The Impact of Information Technology on the Organization of Economic Activity: The 'Move to the Middle' Hypothesis", pp. 19, 22, at <http://portal.acm.org/citation.cfm?id=1189668.1189671> (accessed 21 November 2010).
74. <http://en.wikipedia.org/wiki/Collaborative_software> (accessed 27 December 2009).
75. "Collaborative Tool — Definition", <http://www.wordiq.com/definition/Collaborative_tool>.
76. Brian Shatterfield, "Web Conferencing Tools: Right for You?", 12 February 2010, at <http://www.techsoup.org/learningcenter/internet/page5975.cfm> (accessed 20 April 2011).
77. See, for instance, "Virtual Meeting Smackdown! 15 Top Web Conferencing Services Compared", at <http://www.socialbrite.org/2011/01/19/comparison-top-web-conferencing-services/>.
78. "Collaborative Tool — Definition".
79. <http://en.wikipedia.org/wiki/Drupal> (accessed 27 December 2009).
80. <http://www.oracle.com/us/products/middleware/beehive/index.htm> (accessed 27 December 2009).
81. <http://en.wikipedia.org/wiki/Enterprise_social_software> (accessed 27 December 2009).
82. Mike Heck, "Enterprise Social Software Spurs Connections", *InfoWorld*, 26 August 2009, at <http://www.infoworld.com/d/applications/enterprise-social-software-spurs-connections-804> (accessed 20 April 2011).
83. Noveck, *WikiGovernment*, p. 151.
84. Ibid.
85. Jonas Tallberg, "The Design of International Institutions: Legitimacy, Effectiveness, and Distribution in Global Governance", p. 1, at

<http://www.statsvet.su.se/English/Research/archive/dii/abridged_ project_proposal.pdf> (accessed 21 November 2010).

86. Volker Rittberger, "Multipartite Institution", paper prepared for the 49th Annual Meeting of the International Studies Association, San Francisco, 26–29 March 2008, at <http://www.uni-tuebingen. de/uni/spi/taps/tap52.pdf> (accessed 21 November 2010).

87. H.E. Dr Susilo Bambang Yudhoyono, president of the Republic of Indonesia, "On Building the ASEAN Community: The Democratic Aspect", on the occasion of the 38th Anniversary of The Association of Southeast Asian Nations, Jakarta, 8 August 2005, p. 4, at <http:// www.kbrisingapura.com/docs/Press_release/8b_agust_2005.pdf> (accessed 20 April 2011).

88. Ibid., p. 3.

89. Ibid., p. 4.

90. Soesastro, "Lessons from ASEAN Institution", p. 27.

91. Yuchengco Center, De La Salle University, "Regional Cooperation for Human Development: The UNESCAP and ASEAN Perspectives", Chapter 4 of "South East Asia Human Development Report 2005", at <http://dirp3.pids.gov.ph/seahdr/Chapter4.pdf> (accessed 20 April 2011).

92. Carolina Hernandez, "The ASEAN ISIS and CSCAP Experience", in *The Second ASEAN Reader*, compiled by Sharon Siddique and Sree Kumar (Singapore: Institute of Southeast Asian Studies, 2003), at <http://books.google.com/books?id=_KA-0OkEbrkC&pg=PA280&lp g=PA280&dq=ASEAN+ISIS+experience&source=bl&ots=xEuChgN IG-&sig=AEgl8j5gg1Y68OrW2peKMTlYhHA&hl=en&ei=KI1FS9dTjIr qA-fEjWU&sa=X&oi=book_result&ct=result&resnum=1&ved=0CAc Q6AEwAA#v=onepage&q=ASEAN%20ISIS%20experience&f=false> (accessed 21 November 2010).

93. Ibid., pp. 281–82.

94. <http://www.aseansec.org/18362.htm> (accessed 29 December 2009).

95. Ortuoste, "'Crafting' the State and ASEAN", p. 23.

96. Asian Forum for Human Rights and Development, "Civil Society

Engagement in the Establishment of ASEAN Human Rights Body Lead by FORUM-ASIA and SAPA Task Force on ASEAN and Human Rights", submission to the Parliament of Australia, Joint Standing Committee on Foreign Affairs, Defence and Trade House of Representative, at <http://www.aph.gov.au/house/committee/jfadt/asia_pacific_hr/subs/Sub%2012%20attach%202.pdf> (accessed 20 April 2011).

97. <http://www.seaca.net/viewSection.php?sID=126> (accessed 21 November 2010).

98. Kavi Chongkittavorn, "ASEAN Charter is One Year Old", originally published in *The Nation* (Thailand) on 14 December 2009, at <http://www.asianewsnet.net/news.php?id=9054&sec=3> (accessed 27 December 2009).

99. <http://www.themalaysianinsider.com/index.php/opinion/breaking-views/45888-asean-charter-at-one-a-thriving-tiger-pup--tommy-koh> (accessed 21 November 2010).

100. <http://en.wikipedia.org/wiki/EParticipation> (accessed 27 December 2009).

101. Committee on the Status of Future of Federal e-Rulemaking, "Achieving the Potential: The Future of Federal e-Rulemaking: A Report to Congress and the President", 2008, p. 3, at <http://ceri.law.cornell.edu/documents/report-web-version.pdf> (accessed 27 December 2009).

102. <http://en.wikipedia.org/wiki/Rulemaking> (accessed 27 December 2009).

103. Ibid.

104. Beth Simone Noveck, *WikiGovernment: How Technology Can Make Government Better, Democracy Stronger, and Citizens More Powerful* (Washington, D.C.: Brookings Institution, 2009).

105. Ibid., p. 7.

106. Japan is even worse off — 400,000 applications annually with 1,358 patent examiners with a backlog of 750,000, cited in Noveck, p. 59.

107. Ibid., p. 62.

108. Patent Office Professional Association Newsletter, cited in Noveck, p. 57.

109. "Here is how Peer to Patent works", <http://www.peertopatent.org/> (accessed 27 December 2009).

110. Noveck, p. 99.

111. Noveck, p. 99.

112. Don Tapscott and Anthony D. Williams, *Wikinomics: How Mass Collaboration Changes Everything* (NY: Portfolio, 2006), p. 72.

113. "Engaging Citizens Online for Better Policy-making", OECD Policy Brief, March 2003, <www.oecd.org/dataoecd/62/23/2501856.pdf>.

114. "New Chair Vietnam Pushes ASEAN Community Plan", *Bangkok Post*, 1 February 2010, at <http://www.bangkokpost.com/news/local/30354/new-chair-vietnam-pushes-asean-community-plan> (accessed 20 April 2011).

115. Moe Thuzar, "ASEAN's Socio-cultural Pillar: A Child with "Special Needs", at <http://www.iseas.edu.sg/aseanstudiescentre/asco56-10.pdf> (accessed 20 April 2011).

116. *Unblocking the Roadblocks*, pp. 6–7.

117. Amita V. Acharya, "Constructing Security and Identity in Southeast Asia", *The Brown Journal of World Affairs* XII, no. 2 (Winter/Spring 2006): 157.

118. Ibid., p. 161.

119. Ibid., p 161.

120. Ibid., p. 159.

121. Benedict Anderson, *Imagined Communities: Reflections on the Origin and Spread of Nationalism*, Philippine Edition (Pasig: Anvil Publishing, 2003).

122. Ibid., pp. 42–43.

123. Steven Levy, "The New Wisdom of the Web", *Newsweek*, 3 April 2006, at <http://www.newsweek.com/id/45976/page/4> (accessed 20 April 2011).

124. Adam Singer, "49 Amazing Social Media, Web 2.0 and Internet Stats", *The Future Buzz*, 12 January 2009, at <http://thefuturebuzz.

com/2009/01/12/social-media-web-20-internet-numbers-stats/>
(accessed 20 April 2011).

125. "Yahoo Study Probes Internet Use in Southeast Asia", *Asia Media Journal*, 3 June 2010, at <http://www.asiamediajournal.com/pressrelease.php?id=1839> (accessed 20 April 2011).

126. Ibid.

127. Neither Senior General Than Shwe (head of state) nor Prime Minister General Thein Sein (head of government) has an FB page.

128. <http://beyondsg.typepad.com/beyondsg/>. George Yeo is also in FaceBook, <http://www.facebook.com/georgeyeopage?ref=search&sid=561560521.3118423460..1#/georgeyeopage?v=info>. He also uploaded a short Christmas greetings in YouTube (wearing a collarless UNICEF t-shirt).

129. <http://beyondsg.typepad.com/beyondsg/2009/04/aborted-summit-in-pattaya.html>.

130. Ibid.

131. Ibid.

132. Nathaniel Whittemore, "Obama's Change.org: Citizen's Briefing Book", Change.org, 12 January 2009, at <http://socialentrepreneurship.change.org/blog/view/obamas_changeorg_citizens_briefing_book> (accessed 20 April 2011).

133. <http://usgovinfo.about.com/b/2009/05/13/what-we-told-obama-the-citizens-briefing-book.htm>.

134. D. M. Boyd and N. B. Ellison, "Social Network Sites: Definition, History, and Scholarship", *Journal of Computer-Mediated Communication* 13, no. 1 (2007): 210–30.

135. Thomas Crampton, "Social Media in China: The Same, but Different", at <http://www.thomascrampton.com/china/social-media-china-business-review/>. See also April Rabkin, "The Facebooks of China", at <http://www.fastcompany.com/magazine/152/the-socialist-networks.html>.

136. Ibid.

137. "Social Networking Habits Vary Considerably Across Asia-Pacific Markets", comScore, 7 April 2010, at <http://comscore.com/Press_

Events/Press_Releases/2010/4/Social_Networking_Across_Asia-Pacific_Markets> (accessed 20 April 2011).
138. Ibid.
139. Jenny Preece and Dianne Maloney-Krichmar, "Online Communities: Design, Theory, and Practice", *Journal of Computer-Mediated Communication* 10, no. 4, at <http://jcmc.indiana.edu/vol10/issue4/preece.html>.
140. Linda Gallant, Gloria Boone, and Austin Heap, "Five Heuristics for Designing and Evaluating Web-based Communities", *First Monday* 12, no. 3 (5 March 2007), at <http://firstmonday.org/htbin/cgiwrap/bin/ojs/index.php/fm/article/view/1626/1541>.
141. Ibid.
142. <http://en.wikipedia.org/wiki/Virtual_world>.
143. <http://en.wikipedia.org/wiki/Second_Life>. Unless otherwise indicated, all subsequent information about Second Life is from this wikipedia article.
144. "How Meeting in Second Life Transformed IBM's Technology Elite into Virtual World Believers", <http://secondlifegrid.net/casestudies/IBM>. All information about the event cited subsequently are from this case study conducted by IBM and Second Life Grid.
145. Judith Staines, "European Parliament Launches Citzalia Game", Culture360.org, 13 August 2010, <http://culture360.org/news/european-parliament-launches-citzalia-game/> (accessed 20 April 2011).
146. Ibid.
147. Tim Guest, *Second Lives: A Journey Through Virtual Worlds* (London: Arrow Books, 2007).
148. Frances S. Berry et al., "Three Traditions of Network Research: What the Public Management Research Agenda Can Learn from Other Research Communities", *Public Administration Review* 64, no. 5 (September/October 2004): 543.
149. Jordan and Schout, *The Coordination of the European Union*, p. 24.

150. Ibid., pp. 23, 25.
151. Ibid.
152. E. H. Klijn and J.F.M. Koppenjan, "'Abstract' — 'Public Management and Policy Networks'", *Public Management Review* 2, no. 2, p. 135.
153. W.J.M. Lickert and J.F.M. Koppenjan, "Public Management and Network Management: An Overview", in *Managing Complex Networks: Strategies for the Public Sector*, by W.J.M. Kickert, E.H. Klijn, and J.F.M. Koppenjan, p. 46, at <http://books.google.com/books?hl=en&lr=&id=lB09cQrna6oC&oi=fnd&pg=PR8&dq=related:1QSMxS5uSBwJ:scholar.google.com/&ots=Tx04nYtZ0e&sig=8iTOK6iYh-ztRJ8ZLjWvonezanQ#v=onepage&q&f=false> (accessed 22 November 2010).
154. Joaquin Herranz, Jr., "Network Management Strategies", Working Paper #2006-01, Daniel Evans School of Public Affairs, University of Washington, April 2006, p. 7, at <http://128.95.208.73/files/EvansWorkingPaper-2006-01.pdf> (accessed 27 December 2009).
155. Ibid.
156. Ibid., p. 8.
157. Ibid., p. 10.
158. Ibid.
159. Ibid., p. 12.
160. Michael McGuire, "Is It Really So Strange? A Critical Look at the 'Network Management is Different from Hierarchical Management' Perspective", p. 17, at <http://teep.tamu.edu/Npmrc/McGuire.pdf> (accessed 20 April 2011).
161. Ibid.
162. "The Chief Information Officer: Mastering the Power of Information", at <http://www.leadershipexpertise.com/resources/CIO%20Current%20and%20Future%20Challenges.pdf> (accessed 27 December 2009).
163. "Public CIOs' Role Shifting to Leader", at <http://www.govtech.

com/pcio/Public-CIOs-Role-Shifting-to-Leader.html> (accessed 22 November 2010).

164. "Governing by Network: CIOs and the New Public Sector", at <http://www.govtech.com/magazines/pcio/Governing-by-Network-CIOs-and-the.html> (accessed 22 November 2010).

165. <http://wordnetweb.princeton.edu/perl/webwn?s=interoperability> (accessed 27 December 2009).

166. *eGovernment Interoperability: Guide* (Bangkok: UNDP APDIP, 2007), p. 5, at <http://www.apdip.net/projects/gif/GIF-Guide.pdf> (accessed 22 November 2010).

167. Ibid., p. 3.

168. *e-Government Interoperability: A Review of Government Interoperability Frameworks in Selected Countries* (Bangkok: UNDP APDIP, 2007), p. 1, at <http://www.apdip.net/projects/gif/GIF-Review.pdf> (accessed 22 November 2010).

169. "European Interoperability Framework Supports Openness", at <http://opensource.com/government/10/12/european-interoperability-framework-supports-openness> (accessed 22 November 2010).

170. <http://blog.webwereld.nl/wp-content/uploads/2009/11/European-Interoperability-Framework-for-European-Public-Services-draft.pdf>, p. 1 (accessed 22 November 2010).

171. Ibid., p. 2. Underscoring in the original.

172. *eGovernment Interoperability: Guide*, p. 23.

173. (Denmark) Ministry of Science,Technology and Innovation, "White Paper on Enterprise Architecture", p. 16, cited in ibid., p. 23.

174. Katja Liimatainen, Martin Hoffmann, and Jukka Heikkila, "Overview of Enterprise Architecture Work in 15 Countries", Finnish Enterprise Architecture Research Project (Helsinki: Ministry of Finance, 2007).

175. <http://www.cio.gov/Documents/fedarch1.pdf>.

176. "Tanks in the Cloud: Computing Services Are Both Bigger and Smaller than Assumed", *The Economist*, 29 December 2010, at <http://www.economist.com/node/17797794?story_id=17797794&fsrc=scn/tw/te/rss/pe> (accessed 20 April 2011).

177. <https://apps.gov/cloud/advantage/information/page.do?BV_UseBV Cookie=Yes&keyName=CLOUD_FAQ#t1-a> (accessed 22 November 2010).

178. "U.S. Government Jumps for Cloud Computing in a Big Way", at <http://www.downloadsquad.com/2009/09/16/u-s-government-jumps-for-cloud-computing-in-a-big-way/> (accessed 22 November 2010).

179. <https://apps.gov/cloud/advantage/main/start_page.do> (accessed 22 November 2010).

180. Aaron Levie, "2011: The Enterprise Resets", 2 January 2011, at <http://techcrunch.com/2011/01/02/2011-enterprise/> (accessed 20 April 2011).

181. Amitav Acharya, "ASEAN's Challenges for the 21st Century", *Brunei Times*, 21 July 2009, at <http://www.bt.com.bn/opinion/2009/07/21/aseans_challenges_for_the_21st_century> (accessed 20 April 2011).

182. Thitinan Pongsudhirak, "ASEAN Charter in Deficit After Two Summits", *Opinion Asia*, 3 November 2009, at <http://opinionasia.com/ASEANCharterdeficit> (accessed 20 April 2011).

183. Manuel Castells, "Materials for an Exploratory Theory of the Network Society", *British Journal of Sociology* 51, no. 1 (January/March 2000): 15.

184. Ibid.

ANNEX 1
ICT IN GOVERNANCE AND COMMUNITY BUILDING IN SOUTHEAST ASIA

Information and Communications Technology (ICT) could become the Association of Southeast Asian Nations' (ASEAN's) most useful tool for achieving its goal of "One Vision, One Identity, One Community". More specifically, ASEAN's success in regional integration and community building is also dependent on its ability to use ICT in the areas of: (1) coordinating policies, (2) deepening participation of ASEAN stakeholders, and (3) creating a regional identity, particularly among its youth.

ICT and Policy Coordination

The ASEAN Charter created new regional bodies to enhance the organization's policy-making and coordination efforts. These newly formed groups include the ASEAN Coordinating Council, three ASEAN Community Councils, the ASEAN Sectoral Ministerial Bodies, and the Committee of Permanent Representatives to ASEAN (CPR).

However, creating more committees does not necessarily lead to greater policy coordination — only to more complex processes and increased transaction costs. A more intensive use of ICT by these new ASEAN bodies, including the ASEAN Secretariat, could facilitate better policy coordination at reduced costs.

At present, ICT use in the ASEAN organization is limited to office productivity tools (email, word processing, and spreadsheets) and a web browser. ASEAN should examine new *software-based conferencing solutions* for audio and web-conferencing to complement existing face-to-face ASEAN meetings. Private sector companies that use these capabilities have achieved increased efficiencies, total cost savings, and environmental footprint reductions. Similar outcomes can be expected if ASEAN use conferencing tools.

Another ICT application that ASEAN can use to respond to the challenges of horizontal and vertical policy coordination is *collaborative software*. Collaborative software enables teamwork by transforming the way documents and "rich media" are used in an organization. Web-based collaborative software, which includes scheduling, project management, online chats and meetings, allows various individuals in multiple locations to work jointly on a single document. Like conferencing, collaborative software can make ASEAN more efficient and even reduce the length and/ or frequency of ASEAN working groups and senior officials and ministerial meetings. Collaborative software would also enable the ASEAN Secretariat to discharge its mandated functions more effectively and efficiently.

Collaborative software can also be employed in ASEAN's external relations, particularly in working with dialogue partners, sectoral dialogue partners, special observers and guests.

ICT and Inclusive Regionalism

ICT can help ASEAN achieve its goal of becoming an inclusive institution of governance. ASEAN stakeholders represent a resource — expertise on wide ranging issues — that is not maximized in ASEAN policy-making. e*Participation* — the use of ICT for enabling appropriate stakeholder participation in decision making — could change this state of affairs.

Other regions are already implementing e*Participation* projects. The European Union is funding the development of initiatives such as enhanced citizen participation in the European Parliament and Non-Government Organization (NGO) participation in environmental policy through ICT. In the United States, the public-private partnership "Peer to Patent" project is seeking to transform the U.S. Patent Office's closed, centralized process and constructing an architecture for open participation of the scientific and technical community. In most of these efforts the goal is to harness "outsider" expertise to develop better policies.

An ASEAN e*Participation* initiative could begin by migrating existing face-to-face dialogues between Ministers/SOMs and the private sector and other stakeholders online. For instance, the SOM's engagement with ASEAN Institutes of Strategic International Studies network (ASEAN-ISIS) could be made deeper through ICT. ASEAN could also consider online engagement with "Entities Associated with ASEAN" — approximately seventy-two organizations that support the purposes and principles of the ASEAN Charter.

ASEAN should also consider the use of Wikis — web pages that anyone with permission can create or edit — to connect with ASEAN stakeholders. Wikis facilitate information exchange within and/or between groups/offices, help overcome problems associated with working from a variety of separate locations, and serve as a knowledge management tool.

Blogs — shared online journals also called web logs — should also be used extensively in ASEAN. Usually, ASEAN reaches out to its constituency through the mass media. This, in effect, gives the media the power to filter ASEAN's messages. An ASEAN "policy blog" produced by the Secretariat would allow ASEAN to discuss recent initiatives and other activities directly with its stakeholders without involving a third party.

Wikis and blogs not only promote inclusive governance by widening participation in ASEAN policy development, but also strengthen people-to-people exchange.

Creating a Regional Identity

ICT can help promote/create a regional identity by enabling the citizens of member states to "imagine" an ASEAN community through social network sites (SNS). SNS:

> are web-based services that allow individuals to (1) construct a public or semi-public profile within a bounded system, (2) articulate a list of other users with whom they share a connection, and (3) view and traverse their list of connections and those made by others within the system.

While most are familiar with English-language sites such as Friendster and Facebook, it is important to note that there are also big non-English SNS such as CyWorld (Korean), Mixi (Japanese), and QQ (Chinese). A number of SNS are also accessible by mobile phones — the most widely-used ICT device in Asia.

SNS can serve as an inexpensive and reliable platform for greater people-to-people exchange in ASEAN. For instance, women entrepreneurs in ASEAN can set up a page in a SNS where they can update themselves on events and other activities. Lawyers, accountants and other professionals could also benefit from ASEAN-focused SNS pages. But the main target for social networking should be ASEAN youth.

Various studies have pointed to the importance of SNS to the young. SNS have become an alternative source of information and entertainment for today's youth. The so-called Net generation does not get its news from newspapers, but from the Internet, particularly SNS, and its members prefer watching YouTube over television. Even email is being supplanted by Facebook and QQ as communications media.

ASEAN needs to harness the power of SNS for its own community building efforts, otherwise it may become irrelevant because it is unknown to its youth. It can build on the Secretariat's ASEAN VOICES initiative. It can also host an island in Second Life where information of ASEAN activities can be made readily available to its youth.

Recommendations for ASEAN

ASEAN leaders recognize the importance of ICT. In their last summit, they created a high level task force on connectivity with a mandate of creating a master plan for regional connectivity that includes ICT. On its part, the ASEAN Telecommunications and IT Ministers Meeting (TELMIN), through its SOM, is preparing an ASEAN ICT Master Plan 2010–15 called "Towards an Empowering and Transformational ICT: Creating an Inclusive, Vibrant and Integrated ASEAN".

These bodies, together with the ASEAN Secretariat, could consider adopting the following recommendation of this study:

- Adopt an appropriate, robust, cost efficient, common *conferencing* and/or *collaborative software* to be used by various ASEAN bodies/meetings;
- Develop an appropriate e*Participation tools/initiatives* to enhance engagement with ASEAN stakeholders in policy development;

- Use policy *blogs* and *wikis* as part of ASEAN's knowledge management and information dissemination strategy; and
- Implement a Social Networking Strategy aimed at ASEAN youth. This strategy should leverage the power and appeal of social networking tools to develop an ASEAN identity among its youth.

But the full promise of these initiatives can only be realized if these recommendations are embedded within an overall, integrated ICT framework and implemented by competent individuals. Hence, there is a need to:

- Appoint a chief information officer (CIO) in the ASEAN Secretariat who will provide leadership for developing and implementing ASEAN eGovernance applications.

The ASEAN CIO should immediately spearhead the following activities:

1. Conduct an ICT Audit of the ASEAN Secretariat;
2. Develop an ICT Competency Standard for ASEAN staff
3. Develop an Interoperability Framework for ASEAN
4. Develop an Information Security Policy for ASEAN
5. Develop an Enterprise Architecture for ASEAN, including the development of an ASEAN Community Cloud.

Finally, the successful use of ICT by ASEAN will not rely on technology alone. It requires that ICT — as a tool for enhanced governance through more effective policy-making and for a renewed community building effort in Southeast Asia — be embraced by ASEAN officials at all levels of the organization.

ANNEX 2
Highlights of the ASEAN 2.0
Roundtable Discussions

Mina C. Peralta

Myanmar RTD

The Myanmar Roundtable Discussion was hosted by the Tun Foundation Bank at The Chatrium Hotel in Yangon on 2 April 2010.

The discussants were U Thein Oo, president of Myanmar Computer Federation, Dr Myo Thant Tyn, president of Myanmar NGO Network, and U Tin Nyo, director general of Basic Education.

Comments from Discussants

The time has come for ASEAN to enhance the use of ICT and for a standardized ICT system to be introduced among ASEAN member countries. Steps should be taken to ensure the readiness of member countries to fulfil the necessary infrastructure requirements that would enable them to use ICT effectively. There was emphasis placed on the use of ICT to promote education by aiding teaching and learning.

Open Forum

Attention should be placed on the use of ICT as a tool to improve the ASEAN process and supplement the conventional modes of operation within ASEAN, such as face-to-face meetings, when possible. It was pointed out that ASEAN has not taken advantage of existing ICT tools, especially now when the cost of applications is not a major concern since these have become less expensive.

Government leaders in Myanmar should be urged to recognize the importance of ICT in fostering regional community building. This would lead to reconsidering the manner by which ICT is used at both state and grass roots levels. It was stressed that the government should learn that ICT diffusion results in democratization as it promotes free flow of information and is becoming increasingly important and powerful in sending messages to large numbers of people. However, participants did not see human capacity development as a problem since ICT has become relevant and, therefore, familiar among ASEAN youth.

Participants
1. Dr U Myint
2. U Aye Lwin
3. Daw Moe Marlar
4. U Than Lwin
5. U Paw Lwin Sein
6. U Tun Ohn
7. U Ohn Kyaw
8. Daw Sein Nwe Aye
9. U Thein Oo
10. U Tin Win Aung
11. U Thaung Tin
12. U Aung Zaw Myint
13. U Khun Oo
14. Daw Wah Wah Htun
15. Daw Yi Yi Myint
16. Daw Kyi Kyi Nyein

17. Daw Chaw Khin Khin
18. Daw Kyi Kyi Sein
19. Daw Phyu Phyu Myint Naing
20. Daw Lynn Lynn Tin Htun
21. Daw Thida Thant
22. U Soe Htut
23. U Min Oo
24. Daw Khin Aye Win
25. U Tin Htoo Khaing
26. Daw Aye Myat Thu
27. Daw Sandi Myint
28. Daw Thuzar Lin
29. Daw Marlar Maung
30. U Myint Oo
31. U Aye Kyaw Thu

Singapore RTD

The Singapore Roundtable Discussion was hosted by the ASEAN
Studies Centre of the Institute for Southeast Asian Studies at the
National University of Singapore in Singapore on 5 April 2010.

Open Forum

One of the points raised during the discussions was maximizing
the use of ICT in order to help promote ASEAN and its goals
while it is moving towards economic integration. However,
concerns on the gaps in ICT infrastructure, access, and services
among ASEAN member countries affecting an ASEAN-wide ICT
project were cited. The differing levels of access to ICT and ICT
capacity across ASEAN countries were acknowledged, but were
not taken to mean that an ASEAN-wide ICT project could not be
implemented. In order to implement such a project, ASEAN could

begin by using applications within the organization that do not need high levels of ICT capacity.

There is a need to convince the leaders of ASEAN member countries to develop an ICT-enabled ASEAN organization before ASEAN could begin using ICT more intensively in its operations. This need underscored the importance of a chief information officer (CIO) for the ASEAN organization. There is no position at the ASEAN Secretariat that fulfils this function at the moment. The ASEAN Secretariat might also need to conduct an Information Systems Audit to have an inventory of the organization's ICT assets and their use in furthering the group's goals.

The use of collaborative software and web-based video conferencing was seen as potentially helpful in supplementing face-to-face meetings of ASEAN ministers. ASEAN member countries could make use of these applications regardless of the level of ICT competency of their staff or infrastructure of their country. The positives from the use of collaborative software should be highlighted in order for the targeted users to appreciate the immediate benefits and be encouraged to use it.

A recommendation was made to examine the strengths and weaknesses of Web 2.0 tools in order to maximize their effectiveness in helping promote ASEAN regionalism. In this way, blogs on ASEAN could be used to promote thought leadership, an ASEAN wiki could provide authoritative information on the region, and creating an ASEAN island on Second Life could be used to appeal to ASEAN youth.

The discussion turned to the difficulty of addressing and engaging ASEAN youth who are turned off by current politics. Apart from making use of ICT to involve them in ASEAN, it was suggested that youth-oriented non-government organizations in the ASEAN countries — both local and international — be identified to function as links to encourage youth participation in ASEAN.

Participants
1. Terrence Teo
2. Calvin Wong
3. Maria Ng
4. Vivien Chiam
5. Willy Reyes
6. Enkhbat Dangaasuren
7. Cheryl Soriano
8. Cathy Candano

Indonesia RTD

The Indonesia Roundtable Discussion was hosted by The Habibie Centre in Jakarta on 12 April 2010. The discussants were Director General Djauhari Oratmangun of the ASEAN Cooperation Office of the Ministry of Foreign Affairs, and Ir. Soemitro Rustam, commissioner at the Broadband Network Asia Ltd.

Comments from Discussants
The Indonesian Government recognized the usefulness of collaboration software and social networking, but Indonesian bureaucracy needs to be engaged in order for them to use these ICT applications.

It was disclosed that during the ASEAN meeting in Hanoi, the issue of connectivity, including ICT connectivity, was discussed. During this meeting, it was agreed that ASEAN would develop networking among ASEAN countries under the TelSOM and TelMin processes where TelSOM would implement the policies set by TelMin. The ASEAN Secretariat has appointed a director who would handle matters concerning ICT. The ICT Master Plan for ASEAN for 2011–15 was also discussed in Hanoi.

Open Forum

The digital divide was identified as the primary problem in connecting ASEAN member states and promoting ICT use, with only Malaysia and Singapore ranked in the top 50 of the 2008 United Nations eGovernment Index.

There were concerns raised on how policy on standards in the ASEAN ICT Road Map would be crafted. The issues that surfaced included questions on collaboration in developing and setting up system standards, and the use of free and open source software (FOSS) or proprietary software. It was concluded that ASEAN governments need to come up with the correct policies to stimulate private sector to develop applications and software and engage civil society to use these.

It was noted that although the ASEAN TelMin set up a task force to finance connectivity projects of member countries, there were no provisions in these projects for capacity building of potential users to encourage them to take advantage of the connectivity (training in computer and English literacy) or services for them to access (e*Participation* and eGovernment services).

A suggestion was made regarding tapping environment groups to push for the use of ICT since the use of ICT could be integrated in making the case for reducing the ASEAN organization's carbon footprints.

Participants
1. Nazaruddin Nasution
2. Ramon Sevilla
3. Ramdhan Pohan

4. AM Makka
5. Adi Indrayanto
6. Agung Sarwarna
7. Santhi Serad
8. Marwah Daud Ibrahim
9. Ibrahim Yusuf
10. Aan Premana
11. Yuyun Wahyuningrum
12. Dean Affandi
13. Vita Handayani
14. Rahimah Abdurahim

Brunei Darussalam RTD

The Brunei Darussalam Roundtable Discussion was hosted by the
Brunei Darussalam Institute of Policy and Strategic Studies of the
Ministry of Foreign Affairs at the Rizqun International Hotel in
Bandar Seri Begawan on 14 April 2010.

The discussion was facilitated by Dr Fadzliwati Mohiddin.
The discussants were Lim Hong Beng, Dr Yong Chee Tuan, and
Zailani Hj Abd Malik.

Comments from Discussants

The discussants acknowledged that using collaborative software
would work well in arranging meetings among ASEAN leaders,
but before it could be used by governments of ASEAN member
countries, issues on security and confidentiality need to be
addressed.

The use of collaborative software in policy development was
recognized as a good tool to empower people and have them co-
author policies, but the software that will be used to engage the
citizenry should be evaluated, based on the needs and capacity of

the people who will use it. Including various communities in the ASEAN policy-making process was considered timely since ICT is becoming part of the lives of people in ASEAN. Governments of ASEAN member countries should reach out to communities that have yet to take up or are slow in the use of ICT. Web 2.0 applications could provide a political platform for the government to encourage wider participation at the community level.

The level of comfort of senior government officials using Web 2.0 applications was not considered an insurmountable challenge since there are technological frameworks already available and in use. ASEAN's decision-makers need to use these technologies to establish eGovernance. It was seen that once directives on the use of ICT are issued by senior officials, adoption throughout the organization would happen. Once this occurs, the management of technologies would be the issue and the challenge is for officials to distinguish information that would be involved in policy.

On examining the propositions of ASEAN 2.0, four issues were raised: (1) the promotion of Green ICT, which means lessening ASEAN's carbon footprint; (2) the reduction of costs and expenses in ASEAN, resulting in better budget management; (3) Network generation, examining the future of networks/obsolescence of networks; and (4) the added value it brings to ASEAN. The question raised based on these issues was how to harness ICT solutions to put values up for ASEAN and the next generation of ASEAN citizens while bringing the costs and carbon footprint down. It was determined that a road map is needed to achieve the goals set out by ASEAN with governments learning from commercial companies that use ICT to bring costs down. Engaging the ICT industry in developing content was also recommended, where ICT experts take care of the requirements, while ASEAN minds the policies that guide the use of ICT.

Open Forum

The suggestions made in the ASEAN 2.0 paper may be helpful to the ASEAN High Level Task Force in drafting a master plan on ASEAN connectivity. It was stressed that what was important was that senior government officials understood the technology that would be used for ASEAN 2.0 in order to craft the necessary protocols, procedures, and ICT policies and that these ICT tools work.

Since ASEAN is a network and engages with stakeholders in various sectors, including civil society, it must have good knowledge management. It falls on the governments of ASEAN member countries to synthesize the inputs shared by these organizations, and ICT is needed to do this at both the national and regional levels in order for these ICT tools to be used effectively. Furthermore, user-friendly applications must be available for the citizens, especially the next generation of ASEAN leaders, to opt to use the software. Aside from the use of technology, political issues within the region also need to be examined. These issues may be detrimental in the implementation of an ASEAN-wide pilot programme.

Participants

1. Sheikh Abd Mahdani Sheikh Abd Ghani
2. Baldeep Singh Bhullar
3. Dk Rohayaty Pg Dp Md Yassin
4. Mph Bahrin Shah bin Hj Awg Besar
5. Hj Zaini bin Hj Pungut
6. Dr Fadzliwati Mohiddin
7. Juraidah Hj Musa
8. Dr Hjh Syamimi Haji Mohd Aarif Lim
9. Dr Yong Chee Tuan

10. Wallace Koh
11. Tina Lim Keasberry
12. Zailani Hj Abd Malik
13. Syed Ahmad Fauwaz
14. Pg Hj Abd Rahman PSI Pg Hj Ismail
15. Pg Sarimah Pg Hj Abd Latiff
16. Peter Lee
17. S. Kumaraguru
18. Abd Ghani Pg Hj Metusini
19. Radin Sufri Basiumi
20. Tay Eue Kam

The Philippines RTD

The Philippines Roundtable Discussion was hosted by the Foreign Service Institute of the Department of Foreign Affairs at the Filipinas Heritage Library in Makati City on 19 April 2010.

Ambassador Laura Del Rosario of the Foreign Service Institute opened the discussions. The discussants were Dr Wilfrido Villacorta, Professor Emeritus, De La Salle University and former deputy secretary general of ASEAN, and Carlo Subido, business development manager, Intel Technology — Philippines.

Comments from Discussants

The discussion opened with the information that the ASEAN Secretariat has a committee in charge of ICT, whose function is to provide new computer facilities for the organization, but not to address the issues presented in this study.

The discussants expressed the need for the ASEAN Secretariat and members of national secretariats to use Web 2.0 technologies in managing and coordinating the paperwork and the schedules of its officials. It was noted that there are around 700 ASEAN meetings to be attended by the Secretariat's senior staff alone and there is not enough time to attend all these meetings. The use of ICT in these circumstances could put a cap on the meetings and enable staff to better service these meetings.

With the enactment of the ASEAN Charter, concerns regarding the need to expedite intra-ASEAN transactions and how the ASEAN Secretariat could sustain the necessary vertical and horizontal policy coordination were identified. The need to craft and place communications policies to be followed by the newly created ASEAN institutions was stressed since failure to do so may cause confusion and slow the coordination process further.

The post of the ASEAN CIO needs to be filled if ASEAN is to use ICT effectively to attain its goals. The CIO should be a person who understands the policy direction of the organization and has a deep knowledge of technology so he/she could be helpful. Private companies that have successfully used ICT have CIOs who report directly to the chief executive officer. This may work well in governments and international organizations as well. Since there are still impenetrable silos within ASEAN and decisions made by ASEAN senior officials are hardly translated into domestic laws and policies of member states, the CIO could keep track of compliance among member countries on ASEAN declarations.

Another challenge in using ICT in ASEAN involves the development and agreement on standards (and platforms) among member countries. Should ASEAN agree to use video conferencing more extensively, the organization will have to choose from a variety of standards and platforms currently available. There should also be an agreement on the common standard or platform to use. This further highlights the need for a CIO. The importance of training in the use of ICT was also underscored.

Open Forum

A concern raised in connection with the possible conduct of a pilot project across all member countries was the digital divide in ASEAN. In response, it was noted that while a digital divide in ASEAN exists, ICT use in ASEAN countries has also increased over time. In the specific case of a pilot project, the more important factors to consider are the improvement in bandwidth within government agencies, the ministries of foreign affairs in particular, and human capacity development. A successful project will affect an increase in the use of ICT within the ASEAN organization. The creation of a virtual library for ASEAN-related content, making documents available to various levels of officials at the ministries was suggested as a pilot project.

Issues on security and how to address it were brought up. It was pointed out that most security problems, particularly in the private sector, were due to the people who use the network and not the lack of, or poor, technology. Incidents where individuals shared their passwords, used weak passwords, and/or were not careful in storing passwords were cited. There is security software available to protect data, but in order to maintain security in the overall use of ICT, human capacity development is critical.

In order to develop a concept of ASEAN citizenship, the ASEAN organization should play a facilitating role to foster a sense of community and identity in the region. The organization could make use of ICT to bridge or link various organizations or individuals who have common interests. It could also use blogs and wikis to promote its own activities and projects directly to the citizens of the ten member countries. These activities are needed since they form a basis for imagining "ASEAN" and help build a common identity.

As for the relationship between ICT and democracy, it was pointed out that some ASEAN member countries might not be eager to embrace ICT because it promotes democracy. Governments would most likely take a "risk management approach" to the use of technology where they would, through policy, try and lessen the challenges posed by ICT, and improve the benefits they could derive from it. It was clarified that the ASEAN 2.0 study focuses mainly on how ICT could improve efficiency and effectiveness of the intergovernmental organization.

Though the study only addresses the level of eGovernment, there is a sense that inevitably ASEAN must move towards open governance/multistakeholder governance and eGovernance should also be discussed. The study stresses enhancing and deepening the participation of "Entities Associated with ASEAN", groups and individuals who are already engaged in the process; broadening participation, engaging groups not part of the process, is not a central issue in the ASEAN 2.0 agenda. However, enhancing and deepening the participation of stakeholders also furthers the cause of democratic governance.

The state and direction of ICT use in the Philippines were also highlighted as an area where improvement is needed. The use of ICT in the Department of Foreign Affairs (DFA) was mentioned and it was suggested that a specific session on ICT and the DFA might be worth considering.

Participants
1. Ambassador George Reyes
2. Ambassador Cristina Ortega
3. Fatima Guzman
4. Rachelle Anne Oronce

5. Generoso Calonge
6. Maria Charmaine Guevara
7. Andre Estanislao
8. Francis Maynard Maleon
9. Jan Sherwin Wenceslao
10. Khrystina Corpuz
11. Sharon Johnette Agduma
12. Rhodora Joaquin
13. Luningning Camoying
14. Julio Amador
15. Ariel Bacol
16. Virgemarie Salazar
17. Paul Tajon
18. Yvonne Flores
19. Alana Ramos
20. Ela Neri
21. Dr Rachel Roxas
22. Sherwin Ona
23. France Bi Comeng
24. Armela Razo
25. Al Alegre
26. Maria Juanita Macapagal
27. Dr Grace Jamon
28. May Jean Narne
29. Mary Joy Balayan
30. Karen Kreez Tangco
31. Ma Kathreena Del Rosario
32. Dr Erwin Alampay

Thailand RTD

The Thailand Roundtable Discussion was hosted by the Thai Ministry of Foreign Affairs in Bangkok on 30 April 2010.

Welcome remarks were given by Suchart Saesiengthong, director of Division 3 ASEAN Department of the Thai Ministry of Foreign Affairs, and Dr Chadamas Thuvasethakul, deputy executive director of the National Electronics and Computer Technology Centre (NECTEC) of the National Science and Technology Development Agency.

The discussants were Suchart Saesiengthong, Direk Charoenphol, expert to the National Telecommunications Commission, and Manoo Ordeedolchest, former president of the Association of Thai Computer Industry, ex-director of Software Industry Promotion Agency (SIPA), and an adviser on Information Technology, Sripathum University.

Comments from Discussants

The issue of connectivity was discussed and the discussants pointed out that ASEAN has formed a High Level Task Force on Connectivity that tackles five areas — rails, roads, sea, air, and ICT.

It was acknowledged that ICT and connectivity might help build an ASEAN community by 2015 by helping form an integrated ASEAN, thereby transforming the region economically and socially. Presently, ASEAN is undertaking projects aimed at improving connectivity within the region, such as the ASEAN broadband corridor, which is intended to connect ASEAN member countries, and the ASEAN Internet Exchange, which is set to facilitate the exchange of Internet traffic within ASEAN.

Three parties were identified as crucial to using ICT to make the ASEAN organization work better: the ASEAN member countries, the ASEAN Secretariat and other ASEAN institutions, and ASEAN's dialogue partners. It was further suggested that ICT also be used to redesign work in areas in the ASEAN organization that have been less than efficient. The use of ICT was also seen as possibly re-engineering the cost of the work of the ASEAN Committee on Science and Technology.

ASEAN as a community could make use of ICT to address emerging issues such as disaster management, diseases, climate change, youth and elderly, tourism and — when the year 2015 comes — the mobility of people. ICT could help provide an enabling environment in addressing these issues. The organization should be mindful of the negative impacts of ASEAN connectivity that need to be addressed once it is in place, for example, people smuggling, drug trafficking, environment, and health problems.

ICT could be used as a tool to address the structural issues faced by the ASEAN communities properly. The economic community could make use of ICT to facilitate trade, eCommerce, private-public participation, in terms of the economic community. It was recognized that ASEAN needs to put more emphasis on socio-cultural aspects such as person-to-person interaction, tourism promotion, and sharing of culture. The socio-cultural community could use ICT to promote society and effectively network its people, promote research network (education, infrastructure, supporting possible initiatives such as an ASEAN eLibrary, eYouth, ePolling system), and help national ASEAN secretariats complement the ASEAN Secretariat. It was stressed that education systems within the region should be tapped to promote ASEAN awareness.

In ICT, to promote collaboration and participation, it is the applications used that will enable integration in ASEAN. Since an ASEAN integration plan is similar to undertaking a private sector project, a cue from the private sector should be taken and emphasis be placed on encouraging knowledge-based productivity and serving the individual.

The only way to implement the use of ICT is through the CIO. This function should be emphasized with each member nation having an appointed CIO who will promote ASEAN interoperability and standards and look after the promotion and creation of a new culture of participation.

The areas emphasized in the ASEAN 2.0 paper could provide a platform for people at the grass roots level to participate and co-produce information with the government.

Open Forum

There were suggestions of presenting the results of the discussion to the High Level Task Force on Connectivity so that it might be considered as input in the ASEAN ICT Master Plan. The need for an ASEAN CIO will be highlighted once the initiatives of the High Level Task Force on Connectivity are implemented. Sectoral meetings, such as the TelMin and TelSOM, could start undertaking some of the suggestions in the study since they are the natural constituents of the ASEAN organization that could demonstrate that ICT could make ASEAN more efficient.

The ASEAN Secretariat was pinpointed as another ideal starting point for regional eGovernance since it would benefit from moving beyond email when coordinating meetings and have better document management. The Secretariat should also have a CIO to look after their ICT needs. Since it is ASEAN's goal to have all ASEAN countries connected through ICT, an ASEAN scorecard should be developed so that the member countries are aware of their commitments.

The use of ICT will be driven more by government agencies and therefore governments should be able to pull citizens into using these applications. In this way, it becomes clear that ASEAN is not comprised of governments alone, but also its citizens. Governments of ASEAN member countries should start informing its citizens about collaborative applications so they have an idea on how to participate. Education and human capacity building are important in ensuring the success of the ASEAN 2.0 proposal. ASEAN should become an eLearning society. The human aspect of ASEAN should be emphasized along with the security of its facilities and its programmes. Focus on the positive aspects of ASEAN, such as friendship, community, and helping the disadvantaged, could be integrated in these applications.

It was suggested that the scope of ASEAN 2.0 be defined as an ICT-based information and creative community in ASEAN, and working in ASEAN dialogue. If it works within this definition, ways to promote relationships between ASEAN member countries and its peoples could be found. This should enable people to be more open-minded and enable the next generation of ASEAN citizens to develop the community. Sharing of information among ASEAN member countries could be facilitated with the use of ICT, thereby addressing cultural differences, sharing experiences, and learning languages. However infrastructure development and good governance should not be neglected. A team working on securing software and enhancing cybercrime protection laws should also be created.

It is believed that once ASEAN citizens begin collaborating, it will make ASEAN more competitive. As an example, the Thai ICT sector has a social collaboration project that maps ICT suppliers and they have penetrated the country's various markets. This gives the sector better negotiating power.

In order to find resources easily for projects that may be part of ASEAN 2.0 they could be categorized in three groups:

1)   Projects that need no funding (projects that make use of people and/or community efforts using existing facilities);

2) Projects that may be implemented through public-private partnerships (projects such as the broadband initiatives already being done in the markets; however, the rules and regulations for these projects have to be crafted by ASEAN);

3) Projects that may need ASEAN funding (projects that need to be endorsed by ASEAN leaders and use ASEAN money).

It should be made clear that projects at the community level need no funding at all and private-public partnerships are viable at the national level. However, for the regional level, funding may be secured from ASEAN dialogue partners.

Since the ideas presented in the study are parallel with ASEAN initiatives, the next step is to merge these ideas with ASEAN initiatives.

Participants

1. Pravit Chattalada
2. Professor Prasit Prapinmongkolkarn
3. Rames Sodarat
4. Pishet Durongkaveroj
5. Nattapon Nimmanpatcharin
6. Dr Kanchana Kanchanasuta
7. Pishet Rirkpreecha
8. Saisamorn Narklada
9. Pichaya Juangpanich
10. De Kosol Petchsuwan
11. Pattama Singhara na Ayutthaya
12. Ployrawee Kirkphankul
13. Manu Ordeedolchest

14. Pensri Kantasopatra
15. Dr Nopawan Tanpipat
16. Dr Chadamas Tuwasetthakul
17. Dr Kasititorn Phuparadai
18. Kasama Kongsamak
19. Suwipha Wannasathop
20. Santi Surarat
21. Vajana Chuenthongkham
22. Panuporn Pattarachoke
23. Ajara Panyavanich
24. Parinan Wannasawang
25. Mik Sachapaiboon
26. Apinetra Unakul

Cambodia RTD

The Cambodia Roundtable Discussion was hosted by the
Cambodian Institute for Cooperation and Peace and the
International Relations Institute of Cambodia at the Hotel
Cambodiana in Phnom Penh on 3 May 2010.

Welcome Remarks were given by HRH Norodom Sirivudh of
the Cambodian Institute for Cooperation and Peace, and Dr Neth
Barom, vice-president to the Royal Academy of Cambodia.

The discussant was Ken Chanthan of the ICT Association.

Comments from the Discussant

The discussant suggested that the Cambodian representative to
ASEAN be given a policy brief so that the study's suggestions could
be considered. It was pointed out that groups could lobby the
ASEAN Secretariat to include suggestions made in the study.

The gap in the implementation of ICT-related projects in
ASEAN member countries was a concern. Technologies should be

shared among ASEAN countries in order to develop applications and content before any ASEAN 2.0 project could start. Issues regarding resource mobilization and identifying possible funding donors were raised since these are important for the development and promotion of ICT use in ASEAN.

Open Forum
There should be a policy ensuring universal access to ICT since citizens should have access to communication. Within ASEAN, there should be a body designated to promote ICT, along with applications that could be used by individuals or a community to help them create a cultural identity. The gap in the technological capabilities among ASEAN countries was pointed out, along with the need for a mechanism that will ensure that extensive use of ICT in ASEAN would not disadvantage some member countries.

With the ASEAN High Level Task Force on Connectivity drafting an ASEAN Master Plan for Regional Connectivity that includes ICT, the Task Force is also expected to address the issue of the cost of access to ICT within ASEAN to ensure that access costs in ASEAN member countries do not prohibit citizens from participating in the creation of an ASEAN community. The private sector should also contribute to providing citizens with access to ICT.

The question of whether ASEAN still has the appropriate bodies set up to respond to issues faced by the region in terms of the use of ICT was raised. In order to address the organizational redesign, it was assessed that ASEAN could either hire more people who would help it face the increasing challenges to the region, or work smarter through the use of technology by enabling the people working within the ASEAN organization to do more by using ICT to increase productivity and efficiency.

The four components of the ASEAN region — government, civil society organizations, private sector, and academia — should have their own regional network, aside from the ASEAN network. If these networks were established, their documents and information could be connected through ASEAN and access could be provided to these documents in a sub-regional library in each country.

In trying to achieve ASEAN's goal of regionalism, there is not one solution that would be applicable to all member states since they are in varying stages of development and have varying priorities. What is important is that the member nations come together collectively in committing to the goal of the ethical and productive use of ICT. There should be emphasis on this goal both in the collective and individual action plans of ASEAN countries.

Participants

1. Prince Sisowak Tomiko
2. Heng Bunsong
3. Chim Manavy
4. Hy Sothea
5. Ouk Bonaim
6. Rathavy Annanda
7. Ros Narith
8. Bui Thi Phuong Thao
9. Bui Xuan Hien
10. Keo Phirith
11. Sok Dara
12. Chheng Kimche
13. Hem Sinoch
14. Chea Sang
15. Chhouy Chanry

Lao PDR RTD

The Lao Roundtable Discussion was hosted by the National Authority for Science and Technology — Department of Science and Technology in Vientiane on 5 May 2010.

The discussants were H.E. Mr Humphanh Inthalath, deputy minister and vice-president of the National Authority for Science and Technology, and Somlouay Kittignayon, acting director-general of the National Authority for Science and Technology of the Department of Informatics.

Comments from Discussants

Standards are needed across ASEAN, but creating a common standard might be problematic since there are various authorities handling matters concerning ICT. National security concerns and the free flow of information through ICT are additional challenges that need to be addressed.

Although the proposal looks to benefit ASEAN, it reflected selected realities in the region since most of its suggestions would serve member states with higher levels of ICT advancement more. There is also the issue of human capacity development among ASEAN member countries, an area where Laos is lagging.

Apart from the need to craft ICT standards that would be implemented in ASEAN, Laos needs financial assistance to upgrade its infrastructure and find ways to make the development of ICT sustainable. Suggestions on ways to prepare for the implementation of ASEAN 2.0 for countries such as Laos, Cambodia, and Myanmar, were sought.

Open Forum

During the open forum it was pointed out that Laos recognizes its weakness in ICT. There were questions raised regarding whether it would be better to work on available mechanisms in ASEAN rather than coming up with a new one. The issue of budget was also raised especially since the idea needs to be supported by leaders in ASEAN in order to be implemented.

Participants

1. Anousone Sengphachanh
2. Houmphuk Inthphaipk
3. Sengholour Sengsanang
4. Somsack Inthasone
5. Viengkeo Khaopaseuth
6. Khamsonvanh Phutdavong
7. Senthala Vannsanh
8. Phoutavoy Phoummasak
9. Somsawath
10. Sengtianthr
11. Kham Loi
12. Keovisouk Solapoum
13. Keonarchou
14. Khomsarath Sitthirath
15. Silap Boupha
16. Somphannvanh Sengsourinha
17. Parita Sengtianthr
18. Thanongjunh

Vietnam RTD

The Vietnam Roundtable Discussion was hosted by Vietnam's Institute of Southeast Asian Studies — Vietnam Academy of Social Sciences in Hanoi on 7 May 2010.

The discussants were Dr Nguyen Sy Tuan from the Institute of Southeast Asian Studies, Professor Nguyen Phuong Binh from the Ministry of Foreign Affairs, Tran Kim Hao from the Ministry of Planning and Investment, and Professor Dang Ngoc Dinh from the Centre for Community Support Development Studies.

Comments from Discussants
The study should not only discuss the changes ASEAN needs to make in order to move to ASEAN 2.0 but should also pay attention to the challenges of using ICT to promote an ASEAN community.

In reality, there are many things that still need to be done before Vietnam could fully move to ASEAN 2.0, but the development of Vietnam's economy would enable it to incorporate ICT in its policy-making process. The use of ICT in Vietnam is popular in all sectors and is used in policy-making and development, but if a successful move towards ASEAN 2.0 is to be achieved, the lower level countries in ASEAN should be supported by the higher level countries. This mutual assistance among countries in ASEAN would help promote ICT development and aid in regionalization.

In order to use ICT in horizontal and vertical coordination between and within ASEAN states, the ASEAN organization has to know how to manage these processes since they might have separate points of coordination. When this is determined, ICT could be effectively harnessed in ASEAN cooperation and could allow more non-state actors to participate in ASEAN deliberations. If coordination could be resolved online, it would reduce the number of meetings, thereby allowing ASEAN officials more time to discuss the more important issues during face-to-face meetings.

However, before these online meetings could happen, the security of the network needs to be addressed.

There was concern regarding how all the potential of ICT to promote cooperation and community building in ASEAN could be translated into reality. If these suggestions were to be undertaken by ASEAN, it would need to change its institutional organization since it has no community structure. Furthermore, these changes need to be made based on the manner of interaction among member nations.

It was suggested that community building in ASEAN have both Track 1 and Track 2 diplomacy groups to enable organizations involved in Track 2 activities to become involved in the process. In this manner, academia would have access to policy-makers and have them look at their recommendations and take these into consideration. Furthermore, civil society participation in ASEAN could be accommodated through a web-based venue to receive their comments. However there should be a dedicated unit within the ASEAN organization to manage and plan this process.

Open Forum
The role of ICT in community building is recognized as important, but the challenge lies in using ICT for coordination among ASEAN member countries since there is an ICT development gap within the region and the willingness of ASEAN leaders to commit to using ICT varies. This needs to be addressed in order for ASEAN 2.0 to be successful.

It was acknowledged that there is a gap in the level of technological advancement of ASEAN member countries, but all these countries are already at a certain level of technology that is beyond the rudimentary, which means the minimum level of technology among all the member countries has to be found and it could be the starting point. Successful ICT programmes also require leadership and it should be a requirement that ASEAN leaders buy into the use of ICT and become lead users of technology before telling their citizens to use it.

The use of social networking sites might be problematic in Vietnam. Facebook is blocked in the country because the government does not see its use as positive. Ways should be found to reframe the discussion regarding the use of these sites and make the government see the beneficial side of these social networking sites.

Participants

1. Vo Xuan Vinh
2. Le Phuong Hoa
3. Nguyen Duc Hieu
4. Pham Nhat Minh
5. Nguyen Van Ha
6. Nguyen Thanh Van
7. Vu Cong Quy
8. Tran Le Minh Trang
9. Dam Huy Hoang
10. Tran Khanh
11. Nguyen Ngoc Lan
12. Vu Cong Quy
13. Pham Thanh Tinh
14. Ha Le Huyen
15. Nguyen Duc Hanh
16. Nguyen Thi van Ha
17. Ngo Minh Hai

Malaysian RTD

The Malaysian Roundtable Discussion was hosted by Malaysian ISIS in Kuala Lumpur on 13 May 2010.

Open Forum

The proposal was seen as idealistic in terms of what it wants to achieve since the issues it addresses — access, basic connectivity, and broadband connectivity — have to be addressed across the region. It was suggested that the ASEAN ICT Road Map be consulted against the proposal since the Road Map is the basis of where the resources of ASEAN are being placed in developing an information superhighway in ASEAN.

In Malaysia, the telecommunications industry is already making use of ICT in its processes. Regulators and telecom officials conduct their meetings online to exchange information, documents, and position papers. In this regard, Malaysia could share lessons learned in conducting meetings of less importance online. Their experience found that the degree of success of the endeavour rests on the level of comfort of the user with technology, which is another issue that needs to be addressed, apart from the issues of access, infrastructure and other social issues.

ASEAN needs to start building its own information and content hub since retrieving information from the United States or Europe increases the cost of accessing information and broadband. There are enough people in ASEAN who are looking after the development of infrastructure and that content drives the use of the network. It was emphasized that the proposal is doable, based on the level of infrastructure within the ASEAN member countries at present.

The ASEAN ICT Master Plan charts ASEAN's infrastructure development and should also allow for the content to be produced within ASEAN. In order for Internet traffic to be kept within

ASEAN, its citizens have to be interested in local content. The applications used in ASEAN 2.0 should also be hosted within the region, where one of the ASEAN countries could become the hub for ASEAN information. This initiative should be started since Asians drive content production in new media and the cost of retrieving this information from another continent is restrictive.

The proposal presented solutions, such as the naming of a CIO to ASEAN, which presupposes the soundness of the infrastructure and health of the network around ASEAN. However there are still issues that would be subject to discussion, such as cost and regulatory impediments, limitations on gateways. It is good that there are best practices being presented such as eRulemaking, but developing rules across ASEAN in this manner would be open to sovereignty, jurisdictional issues, and network security, including information laws and harmonization of laws on security.

The examples in the proposal cite examples that could serve as a starting point on how ASEAN could change the way it does things. These examples could serve as benchmarks for involving more people in the process. The issue on network readiness illustrates the fact that leaders should confront problems not just on the supply side, but on the demand side as well. However, in order for these applications to be used at a regional level, the leaders at the ministerial level should be able to appreciate the benefits brought about by ICT.

The promotion of public-private partnership to improve infrastructure within ASEAN was encouraged as this was how Malaysia's Multimedia Corridor came to be. It was suggested that local technologies be considered by ASEAN if it becomes ASEAN 2.0 since it would be the governments of the member economies that would decide which software to use. Rules on local language translation should also be set.

Common concerns shared by ASEAN countries, such as infrastructure and human resource development, and helping the ICT development of the CLMV countries (Cambodia, Laos, Myanmar, and Vietnam) should be addressed and leaders should be committed to the promotion and use of ICT.

Participants
1. Dato Mohammad Sharil Tarmizi
2. Afzal Abdul Rahim
3. Looi Kien Leong
4. Dr Chanrdan Elamvazuthi
5. Dr Amidurin Jorah

ABOUT THE AUTHOR

Emmanuel C. Lallana, Ph.D. is Chief Executive of Ideacorp — an independent, non-profit organization focused on the use of Information and Communications Technology (ICT) in governance and education, in business and the economy, and in transforming society. His work is mainly on ICT Policy Development, Human Capital Development (ICT in Education and Training) and Democratic e-Governance, with focus on m-Governance. He is the project leader of the Pan Asia Network on eGovernance (PANeGOV) — a nine-country research initiative on the use of ICT in good governance and citizen empowerment in Asia, funded by the International Development Research Center (IDRC) of Canada. He is also a consultant of the UN Asia Pacific Training Center for Information and Communication Technology for Development (UN APCICT), based in Incheon, Korea. In this capacity, he developed the module on "ICT for Development (ICTD) Policy, Process and Governance" for the UN APCICT's Academy of ICT Essentials for Asian Leaders. He has also conducted trainings for government officials in India (Hyderabad), Korea (Incheon), Myanmar (Pyon Oo Lin), and Afghanistan (Kabul) on ICT policy-making. On 23 March 2011, Dr Lallana received a letter of commendation from President Benigno S. Aquino III for "laudable work" as a member of the Philippines' High Level Task Force on ASEAN Connectivity.